Claude Simon

Claude Simon, winner of the Nobel Prize for Literature in 1985, is one of the most important French novelists to have emerged since the 1950s. First seen, with Alain Robbe-Grillet and Nathalie Sarraute, as one of the stars of the 'nouveau roman', his distinctive treatment of time, memory and history have made him the Proust of the late twentieth century.

In this new study, Alastair Duncan proposes a new reading of Simon's work based on the premiss that his novels are as much written adventures as adventures in language. Special attention is paid to the major novels of the 1980s, *The Georgics* and *The Acacia*. Simon's development is set in the context of the intellectual and critical debates in which his novels were written and first read, from Jean Ricardou's formalism to post-structuralism, intertextuality and psychoanalytic theory.

Alastair Duncan is Senior Lecturer in French at the University of Stirling. He has written widely on Simon and is the editor of *Claude Simon: New Directions* (1985).

Claude Simon

Adventures in words

Alastair Duncan

Manchester University Press
Manchester and New York
Distributed exclusively in the USA and Canada by St. Martins Press

Published by Manchester University Press
Oxford Road, Manchester M13 9PL, UK
and Room 400, 175 Fifth Avenue, New York, NY 10010, USA

Distributed exclusively in the USA and Canada
by St. Martin's Press, Inc., 175 Fifth Avenue, New York,
NY 10010, USA

British Library Cataloguing-in-Publication Data
A catalogue record for this book is available from the British Library

Library of Congress Cataloging-in-Publication Data
Duncan, Alastair B.
Claude Simon : adventures in words / Alastair Duncan.
p. cm.
Includes bibliographical references and index.
ISBN 0–7190–3650–X
1. Simon, Claude—Criticism and interpretation. I. Title.
PQ2637.I547Z643 1994
843′.914—dc20 93–47155

ISBN 0 7190 3650 X *hardback*

Photoset in Linotron Sabon
by Northern Phototypesetting Co. Ltd, Bolton

Printed in Great Britain
by Bookcraft (Bath) Ltd.

Contents

Acknowledgements

Some of the material in this book originally appeared in *Critique*, *Romance Studies*, *La Revue des sciences humaines* and *Claude Simon: New Directions*. I wish to thank the respective editors and publishers for permission to re-use this material in amended form. I am also grateful to Jérôme Lindon of the Editions de Minuit for permission to quote Simon's novels in French (the English translations are my own); and to *Le Figaro* for permission to quote from an interview with Simon. For grants which enabled me to carry out some of the research for this book in France, I must thank the University of Stirling, the Carnegie Trust for the Universities of Scotland, and the British Academy.

I wish specially to express my gratitude to Claude Simon, both for the manuscript page of *L'Acacia* used on the front cover, and in particular for the interest he has taken in this work. His stimulating comments on drafts of the Introduction and of Chapters 2, 5 and 7 have often made me think again and have saved me from committing some errors of fact. Any remaining errors are of course my own responsibility; my interpretations and conclusions are in no sense sanctioned by Simon.

Finally, my thanks go to Alison Cooper who helped prepare the manuscript, to Anita Roy who has been an alert and supportive editor, and to Catriona who has had to bear my absences, literal and metaphorical, while I have, endlessly it seemed, been finishing this book.

Stirling
October 1993

Abbreviations

The following abbreviations for Simon's works are used throughout. All references are to the first Minuit or Calmann-Lévy editions.

LT	*Le Tricheur* (1945)
CR	*La Corde raide* (1947)
G	*Gulliver* (1952)
SP	*Le Sacre du printemps* (1954)
V	*Le Vent* (1957)
L'H	*L'Herbe* (1958)
RF	*La Route des Flandres* (1960)
P	*Le Palace* (1962)
H	*Histoire* (1967)
BP	*La Bataille de Pharsale* (1969)
OA	*Orion aveugle* (1970)
CC	*Les Corps conducteurs* (1971)
T	*Triptyque* (1973)
LC	*Leçon de choses* (1976)
LesG	*Les Géorgiques* (1981)
DS	*Discours de Stockholm* (1986)
L'I	*L'Invitation* (1987)
L'A	*L'Acacia* (1989)

Introduction

> Dans mes diverses aventures que j'ai évoquées tout à l'heure (révolution, guerre, évasion, maladie, voyages, etc.) j'en ai oublié une qui est pourtant, à mon avis, la principale. J'ai écrit des livres. Et ça, croyez, pour une aventure, c'en est une![1]
>
> Claude Simon

Since Claude Simon was awarded the Nobel Prize for literature in 1985, critical interest in his work has quickened, not least in the English-speaking world.[2] What more is there to say? The primary aim of this book is to introduce Simon to new readers. From various angles I want to show what Simon's novels are about and how they developed from the 1950s to the 1980s. I also intend to place his work in context, partly the context of his life, but more importantly the intellectual and critical climate in which his novels were written and read. Readers already familiar with Simon and his critics will also, I hope, find something to interest them in this book. I place more weight than previous studies on the main novels of the 1980s. In retrospect, I observe, almost with surprise, that the various critical approaches I have adopted seem to have fused and produced something which could be called a thesis or, perhaps more accurately, my own fiction of Simon's fiction. My purpose here is to sketch a background to Simon's works before describing the organisation of the book.

Simon's early life was interwoven into the great events of twentieth-century European history. He was born on 10 October 1913 in Tananarive on Madagascar where his father was stationed as an officer in a colonial regiment. In May 1914 the family returned to France; Captain Louis Antoine Simon was killed in action on 27 August of the same year. It was left to Claude's mother, Suzanne Denamiel, to bring him up in the strongly Catholic atmosphere of her family home in Perpignan, until she too died, after a long illness, when her son was eleven. As a young man, Simon reacted against the aristocratic values of his mother's family. In 1936, at the height of the Spanish Civil War, his anarchist sympathies led him into helping the Spanish Republicans smuggle a shipment of arms from France to Spain. That same summer, he briefly became a member of the Communist Party in order to facilitate the journey to Barcelona. Although Simon has remained liberal in his social and political views, his disillusion with politics and in particular with the hope of revolutionary change seems to date partly from this experience, partly too from his visit to the Soviet Union in the following year.

In August 1939 Simon was called up and rejoined, as a corporal, the cavalry regiment in which he had performed his military service four years previously. After a winter of relative inactivity, his regiment advanced on horseback into Belgium in May 1940, only to retreat in disorder before the motorised armour of the enemy. Simon was captured, dispatched with his fellow prisoners to Germany, then repatriated to a prisoner-of-war camp in Western France from which he immediately escaped. By November 1940 he was back in Perpignan.

Simon did not find his way easily into the other great adventure of his life, the writing of novels. His mother would have liked him to be a priest or a soldier. In practice the intention emerged that he should become a naval engineer. He attended the Collège Stanislas, the most prestigious of private Catholic schools in Paris, and from there went to the Lycée Saint-Louis to prepare for the competitive entrance exam to the *Ecole Navale*; but the discipline of the *lycée* proved irksome to him and although he sat and passed the mathematical *baccalauréat*, he was expelled from the school for absenting himself once too often without leave. Subsequently, he persuaded his guardian to give him an allowance to study painting with André Lhote in Paris. At the same time Simon began to try his hand at photography; some of his photographs were published in professional magazines. This latter talent did not come to public knowledge until the later 1980s. Since then, some of his photographs from before and after the war have been publicly exhibited and published.[3] Although Simon finally abandoned his ambitions as a painter, many of his novels include descriptions of paintings, real and imagined, round which he weaves strands of fiction. Indeed, his visual sense is so strong that he sometimes seems to be fighting against the medium he uses, trying to render simultaneously what in words can only be shown in succession.[4]

Simon has chosen to record that his first attempt at writing for its own sake was descriptive, undertaken during his military service as he supervised soldiers bathing one Sunday at Lunéville in 1935.[5] He has also spoken of two novels, written and torn up, before he was mobilised in 1939. But it was not until after his return to Perpignan in 1940 that he began to devote himself singlemindedly to writing. He had inherited sufficient means from his mother to enable him to do so without having to make a living from his novels. Simon has never had to write for a wide public and his novels have never been printed in large numbers or reached as many readers as they deserve.

The first decade after the Liberation was a difficult time for Simon. In 1951 he contracted tuberculosis which was diagnosed only just in time to save his life. He lay bedridden for many months. Simon has commented of this episode how it sharpened his appreciation of the sights and sounds of the everyday, glimpsed through his bedroom window. But the immediate post-war years did not provide a climate favourable to the development of a writer with Simon's gifts and inclinations. His first novel, *Le Tricheur*, completed in 1941, published in 1945, had received some encouraging reviews. The critic Maurice Nadeau compared it to Camus's *L'Etranger*, written at about the same time but published in 1942. In the years following 1945, however, such stories of young men's discovery of the absurd were going out of favour. Camus's involvement in the Resistance and in political journalism led him to reflect in *La Peste* (1947) on the need for collective action and the limits of its effectiveness. The intellectual and literary scene was dominated by Sartre and by the prestige of the Communist Party. In 1948 in *Qu'est-ce que la littérature?* Sartre called for novels of social and political commitment which would touch the reader by appealing to his or her freedom. The Communist Party required its novelists to adhere to the doctrine of Socialist Realism; literature was to be an arm in the class struggle.

Simon was profoundly out of sympathy with all versions of *engagement*. His discomfiture is evident in the three works he produced in the immediate post-war years. *La Corde raide* (1947) is a book of memories and reflections, formally disjointed and anguished in tone. In *Gulliver* (1952) Simon set out to write something more in tune with the times, a panoramic novel set in specific social and historical circumstances, a southern French town shortly after the Liberation when old scores were still being settled. Its main characters, however, are in no sense model heroes, but undergo that initiation into disorder suffered by all Simon's early protagonists. *Le Sacre du printemps* (1954) tells two stories of such initiations, one based on Simon's gun-running adventure in 1936, the other set in Paris in the 1950s. This novel is resolutely modernist in form and style; the stories interweave in ways reminiscent of Faulkner's *The Sound and the Fury*. Of Simon's four early works, two – *Le Tricheur* and *La Corde raide* – have long been out of print, *Gulliver* he disowns entirely, *Le Sacre du printemps* is still clearly an apprentice work. Their variety shows him to be novelist of ambition but uncertain as to where he is going.

In 1957 Simon found in Jérôme Lindon of the Editions de Minuit a new and sympathetic publisher with whom he has remained ever since.[6] From *Le Vent* in 1957, each of Simon's novels has grown out of his previous work, sometimes almost literally so, since he often takes up and refashions for a new context material originally written for an earlier novel. More and more his study has become a workshop of innumerable fragments waiting to be organised into new patterns. All of his novels are transitional, though some more so than others, since the nature of the patterns has changed over the years. Following *Le Vent*, Simon published three novels in quick succession: *L'Herbe* (1958), *La Route des Flandres* (1960) and *Le Palace* (1962), and then, in 1967, *Histoire*. All of these novels are based to a greater or lesser extent on Simon's experiences – as a child, in Spain, in the Second World War – or on the lives of members of his family. They question what we can know about others or the past since all knowledge is partial, memories fragmentary. Shadowy narrators recontruct events which, though vivid in detail, fade ito one another at the edges, erasing the contours of plot, character and chronological time.

Between 1969 and 1976 Simon published four novels which can to some extent be seen as a group. The first of these, *La Bataille de Pharsale*, is closely related to *Histoire* – it shares some of the same characters and situations – and yet it is very different. It begins from a description which seems to engender what follows. The stories and scenes evoked in *La Bataille de Pharsale* cannot be situated in the mind of a single consciousness. These features are typical of the succeeding novels, *Les Corps conducteurs* (1971), *Triptyque* (1973) and *Leçon de choses* (1976). In these novels Simon uses the present tense; from simple beginnings he elaborates the associations of key words and situations; he interweaves parallel stories. The novels seem to take shape during the time it takes to read them. Finally, in the 1980s Simon published three novels, *Les Géorgiques* in 1981, *L'Invitation* in 1987, and *L'Acacia* in 1989. Of these, *Les Géorgiques* and *L'Acacia* are conceived on a grand scale. They combine features of the two main earlier phases of his work. They give fictional form to Simon's past and the history of his family stretching back to Revolutionary times. They raise questions about history and the writing of history. They set up and develop parallels, echoes and resonances between contrasted stories.

From the later 1950s, the intellectual and critical climate became more favourable to Simon. Jérôme Lindon had made his name as an avant-garde publisher by taking the courageous decision early in the 1950s to publish Samuel Beckett's first French novels. By the time Lindon sought Simon out, he and the novelist Alain Robbe-Grillet, who had become a reader for the firm, had devised a strategy: they would establish a readership for the kind of novels which interested them by promoting the novelists as a group. Robbe-Grillet was a tireless publicist and polemicist. While denying that he was a theorist of the novel, he expounded his views of the novel with persuasive, provocative eloquence in letters to reviewers, in news magazines and in literary reviews.[7] The principal novelists whom Robbe-Grillet defended and promoted were Nathalie Sarraute, Michel Butor, Claude Simon, and, naturally, himself. Nathalie Sarraute was a particularly helpful ally. In 1956 she published a collection of essays, *L'Ere du soupçon*, in defence of her own novels but couched, like Robbe-Grillet's essays, in general terms. Robbe-Grillet reviewed *L'Ere du soupçon* in glowing fashion.[8] By playing down the differences between himself and Sarraute he was able to enhance the perception that a collective movement was under way. In 1957, thanks to a review in *Le Monde* of Sarraute's *Tropismes* and Robbe-Grillet's *La Jalousie*, the putative movement acquired a name: 'le nouveau roman'.[9] The name established the idea of the group. It had the great merit of being vague enough to mean almost anything. But it came with a price. It was so convenient that it encouraged journalists, reviewers and academics to define the novelists primarily in relation to one another and to the theories which their novels, misleadingly, were held to illustrate.

The history of the New Novel can very schematically be divided into three phases. In the first phase, from the mid-1950s to the mid-1960s, Robbe-Grillet was the leading light. The novelists were alike in what they rejected and in their general sense of what the novel should be. They were as opposed to the idea of the committed novel as they were to the continuing realist tradition practised by novelists who regularly carried off the annual literary prizes. Both these types of novel employed unquestioningly forms inherited from the past, in particular character and plot. In that sense they, and not the New Novelists, were formalists. The New Novel, suspicious of character and plot, would self-consciously explore and challenge inherited forms and thus work towards a new realism, psychological in the case of Sarraute, stripped of humanist or absurdist illusions of meaning in the case of Robbe-Grillet. New Novels would be more true to reality in laying bare the processes of their own construction. They would carry forward the tradition of the great modernist novels of the twentieth century: Proust, Virginia Woolf, Joyce, Kafka, Céline, Faulkner.

This first phase shaded into a second, in which the theory developed and hardened. Over this second phase, from the mid-1960s to the mid-1970s, presided the novelist and theorist, Jean Ricardou. Much more than Robbe-Grillet, Ricardou envisaged the New Novelists as a group of like-minded writers who, by agreeing to attend certain conferences, had signalled their intent to work together to advance the practice and theory of the modern novel.[10] The theory which Ricardou systematised continued to emphasise the self-reflexive nature of the novel and its critique of realism, but Ricardou denounced as bourgeois myths the idea that novels represent reality or express the personality of their writers. Such myths present language as something natural, innocent and transparent, whereas New Novels demonstrate the productive power of language and invite readers to reflect critically on language and the myths it fosters.[11]

The idea of the 'nouveau roman' as a group of novelists served all its members by promoting interest in their work. Their novels became well-known to generations of foreign university students, probably better known than they were in France. From the late

1950s to the mid-1980s and beyond the French Ministry of Culture sent New Novelists abroad to represent the French novel; no comparable movement arose to take its place. But while some novelists remained more or less attached to the fringes of the group – Robert Pinget and Claude Ollier both contributed to the 'nouveau roman' colloquium held at Cerisy in Normandy in the summer of 1971 – others reacted against what they perceived to be the misrepresentation of their aims and novels. As early as 1960 Michel Butor changed publishers from Minuit to Gallimard so as to distance his creative works from Robbe-Grillet's theories.[12] Nathalie Sarraute came to the Cerisy conference in 1971, but always insisted, against Ricardou, that her novels aimed to represent an uncharted psychological reality. Robbe-Grillet's disaffection became apparent at the Cerisy colloquium devoted to his novels in 1975. Here he objected that the rigorously schematic character of Ricardou's analyses made his novels reassuringly comprehensible, whereas he had always aimed to produce meanings which were multiple and mobile, and his theoretical essays had knowingly served that aim.[13] Thus, in the long run, emerged a third and final phase of the 'nouveau roman'. Its principal negative characteristic became apparent at a colloquium held in New York in 1982 to which Ricardou was not invited. Robbe-Grillet, Pinget, Sarraute and Simon celebrated Ricardou's absence and distanced themselves from his theories.[14] Their renewed emphasis on the novelist's expression of his or her personality, however, can also be found positively in the quasi-autobiographical nature of some of their works in the later 1970s and 1980s.

Simon has always made it clear that he is a novelist, not a writer with theories about the novel. In the 1950s and 1960s, statements that he made about the novel or his novels were in the main elicited from him by journalists. His views certainly overlapped with those of Robbe-Grillet and Sarraute. In an interview given in 1961, for example, he summed up the ideas which united them as 'refus d'une certaine littérature académique, conventionnelle, refus de toute intrigue, accord – du moins entre Alain Robbe-Grillet et moi-même – sur la non-signification du monde'.[15] Elsewhere he showed common ground with Sarraute in stressing the artifices of the conventional novel and the need for a greater psychological realism, although Simon's concern for memory is very different from Sarraute's interest in fleeting states of consciousness:

> Les gens qui écrivent s'imaginent généralement qu'il faut tout dire, ou plutôt qu'il ne doit pas y avoir de trous. Alors ils remplacent les moments d'absence, qui existent dans la réalité, ceux où ils n'ont rien senti, rien perçu du tout, par une espèce de ciment grisâtre, qui doit faire lien et qui me paraît très faux.[16]

In the 1970s, as Simon's reputation grew, he was increasingly invited to attend conferences and colloquia and he gave a number of lectures mainly about his own work, some of which were later published. They show that at this time Simon shared many of Ricardou's views, although he gives these views his own slant. Similarly, Simon's novels of the 1970s can be interpreted as bearing the mark of Ricardou; and his novels of the 1980s as a sharp rejection of his theoretical mentor. One can also draw parallels between Simon's novels and those of other so -called New Novelists. Over the years, as Ricardou, Robbe-Grillet, Simon and Pinget read one another's novels and exchanged ideas, their novels to a limited extent became more like one another. Simon's novels of the 1970s have perhaps more in common with Robbe-Grillet's of the 1960s and 1970s than is common to their works of the 1950s. Techniques passed from pen to pen, for example the use of the present tense, or neutrally-toned visual description, or the ways in which scenes fade into one another, breaking the illusion of a consistent reality.

Historically, then, Simon's work can be related to the New Novel and shows traces of the New Novelists' common history. But as a dominant perspective in which to view Simon's work as a whole the New Novel is too limiting: it omits too much of what is specific to Simon and it sets him in too narrow a framework. Indeed, part of the fascination of Simon is his sensitivity to a range of influences and to changing times. Although the subject-matter of his fiction is formed largely from the experience of his early life, in theme and form his novels reflect many ideas which have boiled in and beyond France since the 1950s. It is in this wider context that I hope to set his novels.

The plan of this book, then, is doubly chronological. Beginning with the novels of the late 1950s, it broadly follows the order in which Simon's novels were written. It also deals successively with some of the main currents of critical thought which have arguably influenced Simon and have certainly weighed heavily on his critics. In Chapter 1, I try to show how *Le Vent* and *L'Herbe* pointed the way towards *La Route des Flandres*, which forms the centrepiece of this chapter. This is the least self-reflexive of the chapters; it is largely in the mode of phenomenological realism typical of much Simon criticism in the 1960s, although I also dwell on the shift towards an aesthetic of language. In the following three chapters I study and to some extent take issue with three critical perspectives typical of the 1960s and 1970s. Chapter 2 jumps forward from *La Route des Flandres* to *La Bataille de Pharsale*. This chapter is devoted to a critical analysis of Jean Ricardou's contribution to studies of Simon. Chapter 3 proposes a view of what is sometimes called the formalist phase of Simon's fiction, from *La Bataille de Pharsale* to *Leçon de choses*. Broadly speaking, my focus here is on post-structuralism. In Chapter 4, I discuss *Les Géorgiques* in the light of some of the assumptions and implications of theories of intertextuality. In the final three chapters – 5, 6 and 7 – I turn to critical approaches which to some extent gained or regained favour in the later 1970s and 1980s. In Chapters 5 and 6 I step back to take two retrospective views of how Simon's fiction developed. Chapter 5 studies the autobiographical impulse in his work up to *Les Géorgiques*. Chapter 6 uses insights from Freud supplemented by Lacan to offer a psychobiographical interpretation of Simon's development as far as *Leçon de choses*. Finally, Chapter 7 considers *L'Acacia* from the point of view of history and historiography; but, taking my cue from the novel itself, I also try to bring together here, in conclusion, the strands of previous chapters, intertextual, autobiographical and psychobiographical.

This book does not pretend to be a comprehensive study of Simon's novels nor an objective account of these various critical currents. I devote most space to what I see as the peaks of Simon's achievement: *La Route des Flandres*, *Les Géorgiques* and *L'Acacia*. A quizzical observer might claim that I liked Simon but disliked his critics. This is far from being the whole truth, even though I believe that some criticism of Simon has done him a disservice by making his novels seem more difficult to read and enjoy than they need be. My admiration for the logical subtleties of Jean Ricardou's readings of Simon is tempered by scepticism about the framework in which he sets them. To deny absolutely that literature is representational or expressive seems to me untenable. My title acknowledges a debt to Ricardou: 'for us', he wrote, 'a novel is less the writing of an adventure than the adventure of writing'.[17] But for me Simon's novels are written adventures in both these senses, and I place as much emphasis on the first as on the second. Simon's novels deal with the world about us, from the everyday delights of sensuous experience – daybreak, the flight of butterflies – to the great themes of war and history, time and death. They are also about how human beings come to terms with these experiences, with the loss of loved ones and childhood certainties, with the desire for meaning, order, values and origin.

I recognise that both representation and the self are problematic: a novelist does not portray the world as it is or express himself directly in his work; his prose is not a simple instrument or, in Sartre's incautious image, a transparent pane of glass.[18] This puts me to some extent in greater sympathy with critics from the 1970s onwards who have explored the fragmentation of the self in writing, with post-structuralists and Lacanians. Where I part company from such criticism can be illustrated by the epigraph from Rilke's *Duino Elegies* which Simon chose for *Histoire*:

> Cela nous submerge. Nous l'organisons. Cela tombe en morceaux.
> Nous l'organisons de nouveau et tombons nous-mêmes en morceaux.[19]

Criticism, especially in the wake of Derrida, has concentrated on showing how things fall apart. What excites me as much is the opposing force in this drama. The desire for order is both an unconscious need and part of the conscious work of the novelist shaping his novels. I am fascinated and moved by Simon's constantly frustrated, constantly renewed attempts to organise a coherent position, a coherent self in the face of continual dissolution.

1

La Route des Flandres: adventure in words

The subject of *La Route des Flandres* is war, its form is baroque. This tangled road (a 'route de filandres') is vibrant with colour, full of incident, constantly in movement. Swarming figures are frozen in moments of dramatic gesture. Spiralling columns of sentences twist and turn. Fragmentary stories rear up like the stumps of broken arches destined never to meet. This exuberant activity is supported by massive forms which are themselves in tension with one another. Part One of the novel deals mainly with incidents from the autumn of 1939 and the débâcle of May 1940; Part Two tells how Georges, his friend Blum and the jockey Iglésia fared in prison camp; Part Three centres on the night which Georges spends in a hotel bedroom after the war with Corinne, the widow of his commanding officer, Captain de Reixach. Disguising and opposing this chronological framework, Simon sets elements of symmetry which hold time in suspension. At the centre of the novel stands the massacre of Georges's cavalry squadron, flanked on either side by the account of a pre-war race meeting at which de Reixach rode in Iglésia's stead and failed to win. The death of de Reixach, described at the beginning of the novel, recurs frequently, but in most developed form at the very end. Simon once said *La Route des Flandres* has the form of a clover, 'un trajet fait de boucles qui dessinent un trèfle, semblable à celui que peut tracer la main avec une plume sans jamais quitter la surface de la feuille de papier'.[1]

This description aptly combines the symmetrical simultaneity of the visual image with the successive nature of the act of writing. In the novel the central part of the clover leaf, through which the pen passes and repasses, is the dead horse which Georges encounters four times in all (RF, 26–9, 105–7, 241–2, 308). Although these

encounters are placed symmetrically, on each occasion the horse has further changed and decayed.

Neither the symmetry nor the chronology of the novel, however, are immediately apparent. Simon wrote *La Route des Flandres* in fragments which he eventually welded together, matching image to image, theme to theme, seeking contrast and variety as well as continuity. There remain loose ends. Overwhelmed by the luxuriance of detail, some readers fail at first to find patterns in *La Route des Flandres*. They become discouraged; some even give up. Conversely the critic's temptation is to find nothing but patterns, to see settled figures and to forget that *La Route des Flandres* is a kaleidoscope in motion, even although it is the constantly changing shapes which most delight and captivate. One way to impose a fleeting order on these shapes is to view this novel in the context of Simon's immediately preceding work. *La Route des Flandres* fulfils the promise of that work and points to the future in that, more than his previous novels, it is an adventure in words in various senses of that phrase: it tells an exciting story, it explores the resources of language.

From *Le Tricheur* (1945) to *Le Sacre du printemps* (1954), by way of *La Corde raide* (1947) and *Gulliver* (1952), Simon's early works were shot through with the desire to free themselves from conventions – 'truquages' Simon called them in *La Corde raide* – which falsified the representation of reality. *Le Vent* (1957) and *L'Herbe* (1958) went a step further. Stylistically and thematically close to *La Route des Flandres*, these novels continue to pursue realism but simultaneously put it in question. The opening pages of *Le Vent* raise these issues explicitly:

> Et tandis que le notaire me parlait, se relançait encore – peut-être pour la dixième fois – sur cette histoire (ou du moins ce qu'il en savait, lui,

> ou du moins ce qu'il en imaginait, n'ayant eu des événements qui s'étaient déroulés depuis sept mois, comme chacun, comme leurs propres héros, leurs propres acteurs, que cette connaissance fragmentaire, incomplète, faite d'une addition de brèves images, elles-mêmes incomplètement appréhendées par la vision, de paroles, elles-mêmes mal saisies, de sensations, elles-mêmes mal définies, et tout cela vague, plein de trous, de vides, auxquels l'imagination et une approximative logique s'efforçaient de remédier par une suite de hasardeuses déductions [. . .] et maintenant, maintenant que tout est fini, tenter de rapporter, de reconstituer ce qui s'est passé, c'est un peu comme si on essayait de recoller les débris dispersés, incomplets, d'un miroir, s'efforçant maladroitement de les réajuster, n'obtenant qu'un résultat incohérent, dérisoire, idiot. (V, 9–10)[2]

Here questions which had previously been present but secondary in Simon's work now come to the fore, questions about perception, memory, knowledge, the power of reason and the capacity of words to grasp or betray reality. What can be known of others and of the past since all knowledge is partial, perceptions fragmentary, memories selective? As suggested by the passage quoted above and by the subtitle of the novel, 'Tentative de restitution d'un retable baroque',[3] *Le Vent* is not just a story but the attempt to reconstruct that story. The narrator wrestles with fragmentary, biased accounts gleaned at second and third hand, from Montès, the solicitor, a bailiff, a friend of Cécile's fiancé, from witnesses who tail off into the anonymity of 'ils' and 'on'. What memory recalls is the vivid incoherence of perceptions, microscopically accurate, scraped free of the gloss of conventional, conceptualised order. The division of the novel into seventeen brief chapters – more than in any other novel by Simon – accentuates the effect of a narrative fabric 'plein de trous, de vides' (V, 9). Within each chapter, gaps are signalled by recurrent expressions of time – 'puis', 'plus tard', 'un moment après' – and by the pervasive use of 'et', especially in dialogue – 'et lui', 'et elle', 'et Montès' – which suggests the accumulation of discrete pieces of evidence. 'Imagination' and 'approximate logic' prove unreliable allies since they conspire to concoct whatever is plausible: 'Et au fur et à mesure qu'il me racontait la scène, il me semblait la vivre mieux que lui-même, ou du moins pouvoir en reconstituer un schéma sinon conforme à ce qui avait réellement été, en tout cas, à notre incorrigible besoin de raison' (V, 138).[4] To make sense of Montès's story, the narrator draws on literary and cultural models: Montès as the Christlike figure in a baroque altarpiece, or as the central character in a drama by Lope de Vega, Calderón or Ben Johnson; Montès as a Holy Fool, ridiculous, full of goodness and innocence, sowing catastrophe. The narrator switches from model to model; each fits, but imperfectly, successive approximations to elusive truth, as is suggested by the link words and phrases which typify his style: tentative comparisons introduced by 'comme si', 'un peu comme si', and negative statements preceding positive (or slightly less negative) ones, 'non pas . . . mais', 'sinon . . . mais', 'sinon . . . du moins'.[5]

Thus, throughout the novel, the narrative proceeds, each hypothesis demanding revision and refinement, each image summoning another in a constant flood. The tension between the flood of new data and the equally compelling urge to order and control reaches its highest pitch in sections crammed with parentheses and even occasionally parentheses within parentheses. Brackets bear witness at one and the same time to the pressure of extra information and, by their nature as punctuation marks, to the desire to order that information:

> Prétendant même qu'il (ou plutôt son père agissant pour lui, car on ne lui concédait même pas cette sorte de capacité) avait utilisé les derniers restes de prestige – les sabres, les armes damasquinées, les portraits d'ancêtres, la particule, le vieil hótel poussiéreux – pour ravir le consentement, non de la principale intéressée, la jeune fille, mais des parents (d'anciens jardiniers, disait-on, qui avaient arraché de leurs mains à la terre dans l'espace d'une vie à peu près l'équivalent de ce que les généraux sabreurs, leurs descendants et leurs intendants avaient, en cent ans, dilapidé en dettes de jeu, chevaux, femmes, ou plus sûrement encore, dans de mirifiques affaires); appareillant donc (son père) un couple. (V, 110–11)[6]

Here syntax, that 'fade ordonnance, ce ciment bouche-trou' (V, 175),[7] is under pressure. Qualifications, elaborations, additions branch out repeatedly from the trunk of a sentence which, though it begins with a capital letter, lacks the conventional structure of subject, verb and predicate. The absence of measured rhythm, the apparent spontaneity of each new departure give the prose an oral quality. It is as if, setting out to describe a wood, the speaker cannot see the trees for the leaves. In seeking to circumscribe the event in its totality, he is forced into details which set an overall view further and further out of reach. And yet the use of parenthesis ensures that each new surge of detail is marked as a deviation; 'appareillant donc (son père)' is a supreme effort to hold the sentence together, to preserve a sense of order and direction, however precarious. This passage epitomises the inconclusive issue of the narrator's attempt to master the detailed evidence with which he wrestles.

What then is the outcome of the narrator's attempted reconstruction? 'Un résultat incohérent, dérisoire, idiot'? Some perceptive readers of the novel take this view.[8] Seen in the context of Simon's later novels, however, I would argue that the narrator of *Le Vent* achieves relative success. The story is told from a point in time after the events have taken place and nothing in the telling disrupts the basic *donnée*: Montès comes as a stranger to this southern French town to claim the inheritance left him by his father; he becomes enmeshed in events which lead to the death of the woman he loves; eight months later he is forced to sell his inherited vineyard. Although the details of the narrator's interpretation of events and character are hedged with doubt, no alternatives contest them and they receive the blessing of Montès himself. Like the central characters of many of Simon's earlier works, Montès clings fiercely to order and is undone by sexuality; resistant to change, he is rejected by the tight fists and narrow minds of a bourgeois society which sees him as a threat to order and bringer of change, and by the blind force of the wind, symbol of time. While *Le Vent* challenges conventional psychological realism – the solicitor's reduction of human motives to self-interest is attacked with satirical verve – it does so in favour of a Dostoevskian complexity of characterisation handled in Faulknerian mode.[9]

In some respects *L'Herbe* goes further than *Le Vent* in questioning realism. The epigraph picks up the theme that reality cannot be known (and hence cannot be represented): 'Personne ne fait l'histoire, on ne la voit pas, pas plus qu'on ne voit l'herbe pousser.'[10] *L'Herbe* plays on the literal and metaphorical senses of the verb ' to see'. Louise is obsessed by the impossibility of perceiving the passage of time, whether in movement (movement is continuous and perception discontinuous) or in the life of an old woman. Marie is dying, moving from one state to another, yet seems never to have changed, either physically or in the quiet resolution of her actions. She had always had 'le même visage qui maintenant, momifié (ossifié), gisait sur l'oreiller, identique, empreint de cette même invincible expression de paisible virginité' (L'H, 228).[11] Marie fascinates Louise: they form a couple reminiscent of Montès and the narrator in *Le Vent*; but the relationship here is different in ways which attenuate the knowledge which the reader seemed to gain in *Le Vent*. Since Marie is for the most part unconscious, Louise cannot consult or question her. To reconstruct her past, Louise disposes of even more fragmentary remains than the narrator of *Le Vent*: photos, a notebook of domestic accounts. She therefore relies more on imagination. What is more, she lacks the relative lucidity of the narrator and inherits instead the vivid confusion of Montès's perceptions and memories: the ten days of Marie's agony appear to Louise 'non comme une tranche de temps, précise, mesurable et limitée, mais sous l'aspect d'une durée vague, hachurée, faite d'une succession, d'une alternance de trous, de sombres et de clairs' (L'H, 125).[12] *L'Herbe* has the texture which *Le Vent* would have had if narrated by Montès. Incidents come and go in non-chronological succession. Transitions from scene to scene are effected by words which, by their sound or sense, relaunch the narrative in a new direction.

In so far as *L'Herbe* has a plot, it concerns Louise; in so far as character matters, the focus is on Marie and, secondarily, on Sabine, Pierre and Georges. Each of these characters illustrates – heroically, grotesquely or ridiculously – the force of what Simon calls history, and a form of resistance to it, for 'endurer l'Histoire (pas s'y résigner: l'endurer), c'est la faire' (L'H, 36).[13] In the development of this theme the reader may choose to see grounds for the resolution of the plot. Louise fails to leave with her lover perhaps because she has understood only too well the nature of History: a cyclical movement round a motionless epicentre, rendering illusory any hope of pastures new. But the novel resists this explicatory, essentialist reading. Although the three (or four?) meetings between Louise and her lover are told in chronological sequence, the currents and counter-currents of Louise's consciousness make it impossible to discern a pattern of cause and effect: 'personne ne voit l'Histoire'. In fact, the structure of the novel is partly symmetrical: two periods of intense activity frame ten pages of apparent and misleading immobility; partly spiral: the same incidents and motifs recur in modified forms. Nor is it clear when the story is told or even by whom. In successive episodes by Louise to her lover? But the third-person voice which embraces Louise's consciousness uses on occasion the future tense: 'Louise will recall'; and for a third of the novel Sabine replaces Louise as the focusing consciousness. Order is more precarious in *L'Herbe* than in *Le Vent*, things constantly threaten to fall apart. Although *L'Herbe* celebrates Marie's life, treats her family with compassionate irony, arouses sympathy for Louise, the novel reads as if the stories of these lives were not given in advance, remain incomplete, evolve in the telling. Language in *L'Herbe* seems to have its own creative momentum.[14]

Like *Le Vent*, and more explicitly than *L'Herbe*, *La Route des Flandres* is an attempt to reconstruct the past and hence a search for knowledge. 'Il me regarda puis la lettre puis moi de nouveau':[15] the first sentence introduces, with familiar abruptness, a couple reminiscent of the two previous novels, fascinated and fascinator, Georges and de Reixach. As in *L'Herbe*, reconstruction concerns both members of the couple: Georges and his experience of war; de Reixach and his triangular relationship with Corinne and Iglésia. What Simon previously called 'notre incorrigible besoin de raison' (V, 138) demands knowledge of cause and effect. Was de Reixach's death in battle an elegantly disguised act of suicide? If it was, did de Reixach seek to die because he had incompetently led his men into a massacre or because he could no longer bear the shame of his private life? Had Corinne, his young wife, committed adultery with Iglésia, de Reixach's jockey and batman? Although Georges's evidence is even more fragmentary and second-hand than that of the narrator in *Le Vent* – scraps of information from Iglésia, one glimpse of Corinne at a racecourse before the war – *La Route des Flandres* does not begin, unlike *Le Vent*, with explicit premonitions of failure. Doubt however begins to be sown by a series of secondary, parallel stories and by the introduction of a character, Blum, whose function is to question all Georges imagines or believes he knows. The secondary stories reproduce in miniature the themes and enigmas which Georges faces in his experience and in de Reixach's. The general who commanded the army in which Georges and de Reixach fought was believed to have shot himself, unwilling, presumably, to outlive the disgrace of the débâcle. Did that other general who is ancestor to both Georges and de Reixach die a similar death? Georges follows the heroic family tradition in believing that de Reixach committed suicide after a defeat in battle, stripped at last of the optimism, inspired by Rousseau, which had originally encouraged him to throw in his lot with the Revolution. Blum showers Georges with conflicting and mundane hypotheses. Perhaps de Reixach was only cleaning his pistol, or had returned from defeat in Spain to discover a valet in bed with his wife. The story of peasant intrigue is even more opaque. It illustrates once again the disruptive power of sexuality and man's incapacity to resist it: the young peasant is trying to keep out of his house a man who had long since found his way to the woman within. But what precisely are the relationships between the characters? In 1958 Simon had published a version of this story which makes all clear; in *La Route des Flandres* he has taken care to tangle this tale beyond all possible unravelling.[16]

Towards the end of Part Two of *La Route des Flandres*, Simon presents Georges's meeting with Corinne as a last, best hope of attaining certain knowledge.

> Et elle maintenant non plus inventée (comme disait Blum – ou plutôt fabriquée pendant les longs mois de guerre, de captivité, de continence forcée, à partir d'une brève et unique vision un jour de concours hippique, des racontars de Sabine ou des bribes de phrases (elles-mêmes représentant des bribes de réalité), de confidences ou plutôt de grognements à peu près monosyllabiques arrachés à force de patience et de ruse à Iglésia, ou à partir d'encore moins: d'une gravure qui n'existait même pas, d'un portrait peint cent cinquante ans plus tôt . . .), mais telle qu'il pouvait la voir maintenant, réellement devant lui, pour de vrai, puisqu'il pouvait (puisqu'il allait) la toucher. (RF, 230–1)[17]

Part Three of the novel dashes the hope that certainty is within Georges's grasp. Corinne denies that there had ever been any liaison between herself and Iglésia, thus removing the keystone which had supported Georges's speculations about de Reixach. The peasant story peters out in indissoluble confusion: Georges can make nothing of the enigmatic statements of Wack and of the old peasant woman. Blum pours further scorn on Georges's version of his ancestor's story: by describing the sober humdrum environment of his own childhood he affirms that reality is not heroic. Doubt seeps from these peripheral stories to the core of Georges's own experience. The parodied certainty of the maps at headquarters contrasts with the cavalrymen's ignorance of where they were. Was it the Elbe, the Rhine or the Meuse that the prison train crossed in the dark? What really had Georges known or seen of de Reixach's death, his perception of it limited to a single point of view? The last page sweeps away even the apparently indelible, the shining image of de Reixach's anachronistic, absurd, glorious gesture, sabre raised to defy the sniper's bullet: 'l'ai-je vraiment vu ou cru le voir ou tout simplement imaginé après coup ou encore rêvé' (RF, 314).[18]

Thus, progressively, the fiction of *La Route des Flandres* dissolves. Like Georges, the reader flounders

> au milieu de cette espèce de décomposition de tout comme si non pas une armée mais le monde tout entier et non pas seulement dans sa réalité physique mais encore dans la représentation que peut s'en faire l'esprit . . . était en train de se dépiauter se désagreger s'en aller en morceaux en eau en rien. (RF, 16–17)[19]

The past cannot be reconstructed, for how are we to know, 'comment savoir'? Reality is beyond our ken.

More than a failed search for knowledge, however, *La Route des Flandres* is also the account of a personality in dissolution. To convey the 'symbolic death of Georges',[20] Simon expands and refines the techniques to destabilise character with which he had experimented in *Le Vent* and *L'Herbe*.

The social gulf between de Reixach and Iglésia is characterised in their speech. De Reixach's military euphemism – 'vilaine affaire' (RF, 155)[21] – draws a linguistic veil over the massacre of his squadron. Iglésia's status as proud retainer is established by his racy colloquial tone as much as by what he says:

> et lui: 'Reichac vingt dieux t'as pas encore compris: chac l'ixe comme ch-che et le ch à la fin comme k Mince alors jte jure çuilà qu'est-ce qu'il peut être cloche ça fait au moins dix fois que je lui explique t'as donc jamais été aux courses patate c'est pourtant un nom assez connu'. (RF, 46)[22]

The soldiers' banter, set out in single-line paragraphs, has a similarly naturalistic quality, its humour tinged with the desperation of boredom and helplessness (RF, 131–2, 271–3). This in turn is distinguished from the eighteenth-century prose of Georges's ancestor, with its simple sentences (reputedly translated from Italian), pedantic enthusiasm, archaic spelling and unpredictable punctuation (RF, 55–6). Differences and distinctions are not erased in *La Route des Flandres*; hence the novel's strong sense of immediacy, of life evoked in its specific detail. But none of these distinctions is rigid. All the voices which speak at any length in *La Route des Flandres* begin eventually to resemble one another: Iglésia, Corinne, and above all Blum, whom Simon uses to mark this interchangeability. 'Et Georges (à moins que ce ne fût toujours Blum, s'interrompant lui-même, bouffonnant [. . .] et Blum (ou Georges): "C'est fini?", et Georges (ou Blum): "Je pourrais continuer", et Blum (ou Georges): "Alors continue" (RF, 187–8).[23] This melting of voices into one makes Georges the sole guarantor of their reality. Who then is Georges, where does he speak from, and to whom?

A first answer to the last question comes early in the novel: ' "Ouais! . . ." fit Blum (maintenant nous étions couchés dans le noir, c'est-à-dire imbriqués entassés au point de ne pas pouvoir bouger un bras ou une jambe' (RF, 20).[24] Twenty pages later this supposed place of origin – the cattle-truck carrying Georges and Blum to prison camp in Germany – is put in question by references to a hotel bedroom and a new companion; Georges is now reflecting on 'the dead years', 'écoutant le silence, la nuit, la paix, l'imperceptible respiration d'une femme à côté de lui' (RF, 42).[25] Is Georges then telling the whole story to Corinne, or ruminating over it as he lies in bed with her (RF, 95)? The reader who jumps to this conclusion has jumped too quickly. Part Two seems to be told from the prison camp. Even Part Three, where references to the night with Corinne come again to the fore, past tenses continue to mingle paradoxically with adverbs conveying the present: 'maintenant nous étions couchés' (Part One, 20); 'à présent ils se tenaient tous trois' (Part Two, 169); 'pouvant maintenant entendre l'air pénétrer en elle' (Part Three, 269).[26] Thus the reader is hustled from context to context and the character floats ungrounded in space and time.

Georges's identity disintegrates further through Simon's manipulation of point of view. It is not simply that the narrative alternates between first and third person; the disorientating effect is cumulative. The first switch, from first to third person, does not disrupt the description of the decaying horse (RF, 27). In one sense it brings clarification since the as yet anonymous consciousness acquires a name, Georges, thus providing readers with an important piece of the jigsaw with which they are juggling. This new perspective is more or less maintained until half-way through the novel since each time the third person returns it is contained within quotation marks. This safeguard fails however when the horse-race turns into the massacre of the cavalry squadron. At this moment Georges ought to have died, as did almost all his comrades; it becomes in the narrative the moment when his identity splits. From then on, switches between first and third person abandon the pretence of container and contained. The narrative shifts unpredictably, unmotivated by specific causes. Georges is neither self nor other, but suspended between the two. This instability of point of view, like those of time and place, deprives the reader of any sense of a fixed centre, an originating core. We can be sure of nothing except that a voice is speaking: Georges consists only of words.

What then is the power of words? What can they achieve? *L'Herbe* showed a loss of faith in words over three generations. Georges's illiterate peasant grandfather was 'pénétré, imbu d'une superstitieuse confiance dans ces mots qu'il ne pouvait qu'entendre et prononcer (et peut-être en les écorchant)' (L'H, 231).[27] His son, Pierre, a professor of philology, gained mastery over words and access to 'le savoir, la science, ce que renferment les livres' (L'H, 231).[28] He continued to use words, though doubting whether they brought certainty of knowledge. The grandson, Georges, rejects them (with admirable eloquence):

> parce que je voudrais n'avoir jamais lu un livre, jamais touché un livre de ma vie, ne même pas savoir qu'il existe quelque chose qui s'appelle des livres, et même, si possible, ne même pas savoir, c'est-à-dire avoir appris, c'est-à-dire m'être laissé apprendre, avoir été assez idiot pour croire ceux qui m'ont appris que des caractères alignés sur du papier blanc pouvaient signifier quelque chose d'autre que des caractères sur du papier blanc, c'est-à-dire très exactement rien. (L'H, 152)[29]

Education is in question here: can book-learning fulfil its promise to give human beings access to knowledge and hence control over their destinies? Equally and more specifically, the relationship between language and reality is at stake. If words fail to give access to knowledge, is it because language does not function in such a way as to reach reality, or because 'reality' does not exist (or even perhaps because the question is misconceived, the relationship between language and reality not being reducible to such terms)? *L'Herbe* hints at answers to some of these questions, but leaves unexplored the mystery of Georges. Why does he reject so vehemently all that his grandfather and father stood for? *La Route des Flandres*, originally an episode in *L'Herbe* for which there was in the end no room,[30] is in a sense the elaborated answer to that question.

From *La Route des Flandres* Simon excluded Louise and Marie, and cut down Sabine's role while adding to it a dash of dynastic pride. Pierre survives as the decaying satirised representative of the humanist tradition. With wavering faith, he believes that human culture, stored in the written word, can with profit be transmitted from generation to generation to realise human happiness. The experience of Georges gives the lie to this faith. When Pierre bemoans the destruction of the library at Leipzig, Georges replies from prison camp that

> si le contenu des milliers de bouquins de cette irremplaçable bibliothèque avait été précisément impuissant à empêcher que se produisent des choses comme le bombardement qui l'a détruite, je ne voyais pas très bien quelle perte représentait pour l'humanité la disparition sous les bombes au phosphore de ces milliers de bouquins et de papelards manifestement dépourvus de la moindre utilité. Suivait la liste détaillée des valeurs sûres, des objets de première nécessité dont nous avons beaucoup plus besoin ici que de tout le contenu de la célèbre bibliothèque de Leipzig, à savoir: chaussettes, caleçons, lainages, savon, cigarettes, saucisson, chocolat, sucre, conserves, gal . . . . (RF, 224–5)[31]

An aspect of Georges's initiation by war is to discover that he is essentially a creature of basic needs – warmth, food, a minimum of comfort, sex – and that human society, epitomised in the prison camp, is a system of exchange organised to satisfy these needs. The critique of humanism which runs through the novel finds expression in the animal imagery applied to human beings – horse, wolf, goat, monkey, swan – which culminates in the sexual descriptions of Part Three: 'je n'etais plus un homme mais un animal un chien plus qu'un homme une bete' (RF, 292).[32] *La Route des Flandres* belongs, at least in part, to a tradition of literary works issuing from the Second World War – from Vercors's stories and novels to Jorge Semprun's *Le Grand Voyage* – which put humanism in question, to attack or defend it.

In *La Route des Flandres*, however, Simon's suspicion of words goes beyond this thematic debate and the point previously reached in his fiction. The narrator of *Le Vent* distrusted syntax, the capacity of ordered language to mask the nature of reality, 'ce foisonnement désordonné, sans commencement, ni fin, ni ordre' (V, 75).[33] The question in *Le Vent*, therefore, was whether, by liberating words from conventional syntax – by the use of shifting, probing, twisting sentences – the real could be more perfectly rendered. *Le Vent* was the conscious culmination of a continuing effort in Simon's fiction: to bring together themes treated separately in the early novels and to express the complexity of the real: the changeless cycle of History;

the fluidity of consciousness and the vivid confusion of memories. *La Route des Flandres* can be seen as a more perfect realisation of that aim. Georges's disillusion echoes and amplifies that of Simon's early characters. His memories blend confusion with precision in unequalled measure. The stories which he constructs illustrate varieties of disillusion and the changing shapes of History which never changes. But *La Route des Flandres* is not only the fulfilment of Simon's early realist aesthetic; it also marks the point where an accumulating change in the practice of his writing challenges the aesthetic and forces him to abandon it: *La Route des Flandres* dramatises the impossibility of using words primarily as instruments.

Like *L'Herbe* but more so, *La Route des Flandres* gives an impression that much of the momentum of the narrative comes not from outside, from memories demanding to be transcribed but internally, from language itself.[34] It seems that the novel does not re-present a world which exists prior to the text but rather that the world of the fiction is produced by words. A striking example of this self-generation begins on page 18 when, to clarify de Reixach's reaction to events on the Flanders road, a comparison is made with how he might have behaved in different surroundings before the war: 'un peu interdit, impatient, comme si dans un salon quelqu'un l'avait brusquement abordé sans lui avoir été présenté ou interrompu au milieu d'une phrase par une de ses remarques hors de propos (comme par exemple . . .' (RF, 18).[35] This deviation, not untypical of *Le Vent* or *L'Herbe*, continues for three lines before the parenthesis closes and the text returns to the war. Twenty lines later, however, the text deviates again:

> reprenant avec ce petit sous-lieutenant sa paisible conversation du genre de celles que peuvent tenir deux cavaliers chevauchant de compagnie (au manège ou dans la carrière) et où il devait sans doute être question de chevaux, de camarades de promotion, de chasses ou de courses. Et il me semblait y être, voir cela: des ombrages verts avec des femmes en robes de couleurs imprimées. (RF, 19)[36]

Triggered by a comparison and by the words 'chevaux' and 'courses', a whole page now evokes the pre-war aristocracy in typical surroundings, displaying the self-conscious, self-confident formality of their manners. The same scene recurs on page 22. Eventually, the series will culminate in the description of the horse-race which, with the massacre of the squadron, forms the climax to Part Two. It seems then that a narrative cell has split in two: from a simple comparison have germinated horses, jockeys, a social class at play, the complex subversion of its codes of conduct in the behaviour of de Reixach, Corinne and Iglésia.

This productive power of language does not immediately put in question its capacity to capture a pre-existing reality. Despite the tentative comparisons, the lists of synonyms, the expressions of doubt – 'sans doute', 'peut-être', 'tout au moins' – the novel begins optimistically. Although the luxuriant detail of these pre-war scenes belong to a narrative fashioned, so to speak, from its own rib, Georges at this stage has faith in words: 'et cette fois Georges put les voir, exactement comme si lui-même avait été là' (RF, 144).[37] Whereas the narrator of *Le Vent* was suspicious of this facility, seeing it as a sign of construction rather than reconstruction, Georges rejoices in the power of words to make him a witness to such scenes. Once again it is Blum's role (to the extent that he can be distinguished from Georges) to initiate doubt. Why did de Reixach ride in the race rather than leaving it to his jockey?

> Bref: peut-être a-t-il pensé qu'il ferait alors, si l'on peut dire, d'une pierre deux coups, et que s'il parvenait à monter l'une il materait l'autre, ou vice-versa, c'est-à-dire que s'il matait l'une il monterait l'autre aussi victorieusement, c'est-à-dire qu'il l'amènerait elle aussi au poteau, c'est-à-dire que son poteau à lui l'amènerait victorieusement là où il n'avait sans doute jamais réussi à la conduire. (RF, 175)[38]

Blum's scabrous speculations mock Georges by parodying his use of language. Here language is not used as an instrument to capture or convey truth; Blum allows it to run riot, giving words free rein to summon others by *doubles entendres* and by analogies of sound and sense: 'monter'/'mater', 'monter une femme'/'un cheval'. The infinite fertility of language must make one despair of finding truth (since words will apparently say anything) and simultaneously diverts attention from any truth language might convey towards words themselves and their relationships in the text.

Step by step Simon obliges the reader to take seriously the view that all Georges's efforts may amount to no more than a game with words. Engendered from one other, the stories of *La Route des Flandres* share similar motifs, situations and themes. In Parts Two and Three of the novel Simon increasingly uses these analogies, and creates new ones, to slip from one story to another and leave the reader momentarily perplexed. In the following example horses and riders transit from the racecourse to the scene of the squadron's ambush:

> pouvant voir comme s'il n'en avait été qu'à quelques mètres l'encolure de la pouliche couverte d'une écume grise à l'endroit où flottait la rêne, le groupe, le cortège hiératique et médiéval se dirigeant toujours vers le mur de pierre, ayant maintenant traversé l'embranchement du huit, les chevaux de nouveau cachés jusqu'au ventre par les haies de bordure disparaissant à demi de sorte qu'ils avaient l'air coupés à mi-corps le haut seulement dépassant semblant glisser sur le champ de blé vert comme des canards sur l'immobile surface d'une mare je pouvais les voir au fur et à mesure qu'ils tournaient à droite s'engageaient dans le chemin creux lui en tête de la colonne. (RF, 154–5)[39]

This passage illustrates well a paradoxical characteristic of *La Route des Flandres*: a strong sense of reality, of visual presence, and the simultaneous questioning of that effect. When does the transition between the two scenes take place? Only on reaching 'le chemin creux' does the reader realise that he or she has been switched on to a different track. But backtracking from there proves of little help. Where the text branches is not to be found. In fact, everything from 'l'embranchement' to 'le chemin creux' can be read as referring to either the horse-race or the squadron's last ride. Forced to question what the 'reality' is to which the words refer, the reader is led to doubt whether they capture reality at all.

The complexity of the system of internal references depends not solely on such transitions. The passage quoted above, for instance, has strong links with an earlier passage in which Simon describes how de Reixach continued to advance after he had been struck by a sniper's bullet:

> comme ces canards dont on coupe la tête et qui continuent à marcher se sauver parcourant grotesquement plusieurs mètres avant de s'abattre pour de bon: rien qu'une histoire de cous coupés en somme puisque selon la tradition la version la flatteuse légende familiale c'était pour éviter la guillotine que l'autre l'avait fait avait été contraint de le faire Alors ils auraient dû changer leur blason depuis ce jour-là remplacer ces trois colombes par un canard sans tête. (RF, 90)[40]

With typically caustic humour, this passage plays ducks and drakes with de Reixach. The colloquial French for a sniper, 'un canardeur', prompts the image of the captain and his ancestor, the general, as 'canards dont on coupe la tête'. This leads to the burlesque suggestion for a family crest appropriate for the de Reixach family, since both men lost their heads in more senses than one. In the passage on page 154 the severed bodies are at first not those of men but of horses 'coupés à mi-corps'. But this, together with the heraldic connotation of 'sur le champ de blé vert', quickly brings back the ducks. And from there it is a short step to de Reixach 'à la tete de la colonne', the column of horsemen but also the spinal column, from both of which he, as head, will shortly be severed. Further relationships could be traced round the novel. For example the 'haies de bordure' of page 155 abound in this novel without fixed boundaries. The effect, however, is already clear. The more Simon multiplies analogies, the more Georges seems betrayed by language, the words referring not to a given reality but to one another.

Certain passages in Part Three bring to a climax the disillusionment with words which affect Georges and the reader, and complete their initiation into the game of language.

> alors je me jetai par terre mourant de faim pensant Les chevaux en mangent bien pourquoi pas moi j'essayai de m'imaginer me persuader que j'étais un cheval, je gisais mort au fond du fossé dévoré par les fourmis mon corps tout entier se changeant lentement par l'effet d'un myriade de minuscules mutations en une matière insensible et alors ce serait l'herbe qui se nourrirait de moi ma chair engraissant la terre et après tout il n'y aurait pas grand-chose de changé, sinon que je serais simplement de l'autre côté de sa surface comme on passe de l'autre côté d'un miroir où (de cet autre côté) les choses continuaient peut-être à se dérouler symétriquement c'est-à-dire que là-haut elle continuerait à croître toujours indifférente et verte comme dit-on les cheveux continuent à pousser sur les crânes des morts la seule différence étant que je boufferais les pissenlits par la racine bouffant là où elle pisse suant nos corps emperlés exhalant cette âcre et forte odeur de racine, de mandragore, j'avais lu que les naufragés les ermites se nourrissaient de racines de glands et à un moment elle le prit d'abord entre ses lèvres puis tout entier dans sa bouche comme un enfant goulu c'était comme si nous nous buvions l'un l'autre nous désaltérant nous gorgeant nous rassasiant affamés espérant apaiser calmer un peu ma faim j'essayai de la mâcher pensant C'est pareil à de la salade. (RF, 258–9)[41]

Here the text oscillates between the poles of literal and sexual hunger. Analogies are rampant: the play of sound and sense from 'boufferais les pissenlits' to 'bouffant là où elle pisse';[42] the implied pun which leads from 'gland' as acorn to 'gland' as male sexual organ; the ambiguous reference in the phrase 'la mâcher' which could mean 'chew it' (the grass) or 'her' (Corinne). Everything here eats or is eaten: horses, ants, grass, Georges, dandelions, Corinne, shipwrecked men and hermits. Since eating implies transformation of the eaten, these images form part of a more comprehensive series of images of metamorphosis: Georges turning into horse and then to earth is joined in this transformation by Corinne, 'nos corps emperlés exhalant cette âcre et forte odeur de racine, de mandragore'. These two patterns of images can be read as a double commentary, two *mises en abyme* of the construction of *La Route des Flandres*. The different stories, incidents and scenes of the novel, once separate and distinct, have now devoured one another; only the words of which they were formed remain. The images of metamorphosis recall how this mutual consumption has taken place, through the power of words to summon up other words, to transform and retransform themselves in infinitely varied combinations.

Near the beginning of Part Two of *La Route des Flandres*, Georges interprets the look in the eye of a dying horse: 'comme s'il avait abandonné, renoncé au spectacle de ce monde pour retourner son regard, le concentrer sur une vision intérieure plus reposante que l'incessante agitation de la vie, une réalité plus réelle que le réel' (RF, 131).[43] Behind the urgency of the quest on which Georges embarks lies the hope of achieving this vision, a hope of solidity, of a still centre beyond the changing forms of life, of a transcendent truth which, as in Baroque art, will guarantee the reality of appearances. *La Route des Flandres* tells the story of the disappointment of that hope. Georges tries to reconstruct the past; he fails. In retelling his memories, he is seeking to establish a coherent myth of the past on which identity can be founded; and the novel invites the reader to take part in this quest. It ends unsuccessfully. Georges cannot decide between 'il' and 'je'; the reader is progressively compelled to abandon his attempts to impose unity and forced to recognise that Georges is a voice without a source, a coherence always just out of reach. Finally, as the personality of Georges dissolves, another kind of failure becomes apparent: words are seen to compose a vast system of signs summoning one another and referring backwards and forwards to one another in endlessly complex relationships but never reaching outwards to the reality they strive to grasp. At one point Georges comments that he and Blum were using language to console themselves: 'de façon [. . .] à faire surgir les images chatoyantes et lumineuses au moyen de l'éphémère, l'incantatoire magie du langage, des mots inventés dans l'espoir de rendre comestible – comme ces pâtes vaguement sucrées sous lesquelles on dissimule aux enfants les médicaments amers – l'innommable réalité' (RF, 184).[44] Thus the novel itself appears, not as a naming of reality but, like the cry of the cock, 'comme un parodique appendice à la bataille' (RF, 252), written in consciousness of failure and only because, in Pierre's words, 'rien n'est pire que le silence' (RF, 35).[45]

Loss and dissolution are the leitmotifs of *La Route des Flandres*: the physical destruction of war, the inner decay of metaphysical and humanistic certainties. Certainly, the novel is cast in a tragic mould. Nevertheless, there are strong grounds for agreeing with Bernard Pingaud's early, nuanced and contrasting assessment of the work.

> Par un curieux et ultime renversement, ce monde usé par le désastre apparaît finalement plus riche que toutes les formes qu'il détruit. La mort y règne partout; mais elle ne peut empêcher une autre puissance – qu'il faut bien appeler la vie – d'avoir le dernier mot. [. . .] C'est plutôt sa profusion, son lyrisme qui nous écrasent, comme si, délivrés de croire que nous existons, nous découvrions, dans sa ruine même, l'obscure et jubilante confusion de l'existence.[46]

A life-force courses through *La Route des Flandres* which can inspire not depression but exhilaration, and which suggests that, even at this stage, Simon had attained that dual vision of the world which he so admired in Cézanne: 'une connaissance substantielle, dans le même moment désenchantée et éblouie' (CR, 119).[47] *La Route des Flandres* has a superabundance of story and character. It teems with movement and colour. Its imagery unfolds in luxuriant, inexhaustible profusion. Behind all this wealth lies the new freedom which Simon allowed himself in *La Route des Flandres* to follow where language leads. To the extent that *La Route des Flandres* dramatises a crisis in representation, it is a crisis of growth. The discovery that language cannot represent reality is experienced as loss but also, potentially, as liberation. Although words cannot represent 'the essential core of the real' because no such thing exists,[48] they have the power to evoke appearance and they bear the multiple traces of all they have ever referred to. Two minus two leaves nothing. But language does not work like that. The different versions of the death of de Reixach's ancestor ring in the mind, however incompatible. When the points change, the reader is neither at the races nor at an ambush, but somewhere between, simultaneously at both. *La Route des Flandres* is a marvellous exhibition of the power of language to bring together the apparently distant, to overturn established relationships and set others in their place. Through an endless play of associations, settings merge into settings, scenes into scenes, characters turn into one another, human beings become animals, and horses are everywhere in every possible guise and state, as cavalry mounts and thoroughbreds, chargers and centaurs, rearing, racing and jumping, dying, decaying and dead. In no other novel does Simon so strongly convey the feeling that life is a matter of seething energy, anarchic and formless; but *La Route des Flandres* must also be admired for the discreet and subtle patterns by which Simon disciplines anarchy and gives shape to formlessness.

2

Ricardou, Simon and *La Bataille de Pharsale*

From the mid-1960s to the late 1970s, and even beyond, one critic exercised a profound influence on the way Simon's novels were read. Although Jean Ricardou wrote relatively few articles specifically on Simon, and he returned again and again to the same few novels – *La Route des Flandres, La Bataille de Pharsale, Les Corps conducteurs, Triptyque* – his books provided a critical and theoretical framework into which Simon seemed to fit. His dominance was increased by the role he took in the series of colloquia held at Cerisy in Normandy in the early 1970s. Co-chairman and co-editor of the colloquium on the New Novel in 1971, he was sole organiser, chairman and editor of the conference devoted to Simon in 1974.[1] On both occasions, but perhaps particularly on the second, he dominated the proceedings, almost as much by force of personality as by the systematic rigour of his thought. There was no getting round Ricardou, either for those present or for those who subsequently read the proceedings. Two further factors established his place. First, he enjoyed a special relationship with Simon. Simon attended both the colloquia at Cerisy. There and elsewhere, he spoke warmly of how much Ricardou had helped clarify what in his own writing he was doing or aimed to do.[2] Second and most important, he was not alone: his work was a manifestation of more general trends which I shall be considering in this and the next two chapters. It was not Ricardou alone who influenced readings of Simon but the climate of critical thought which he exemplified in a particular form.

Ricardou's theories did not spring fully-formed from the earth: his first notable contribution to Simon studies, 'Un ordre dans la débâcle', 1960, was in part a conventional analysis of the themes of *La Route des Flandres*.[3] His ideas have continued to develop, although he prefers to speak of their widening application rather than change.[4] Nevertheless, from the mid-1960s to the mid-1970s his work was remarkable for its sustained attempt to construct a systematic analysis of the New Novel, promulgated in his contributions to the debates at Cerisy and his three books of the period, *Problèmes du nouveau roman* (1967), *Pour une théorie du nouveau roman* (1971) and *Le Nouveau Roman* (1973). This is the period with which I am concerned here.

While *Problèmes du nouveau roman* and *Pour une théorie du nouveau roman* are collections of essays, *Le Nouveau Roman* is the most complete summary exposition of Ricardou's ideas. In that work, his lapidary definition of the New Novel runs as follows: 'Avec le nouveau roman, le récit est en procès: il subit à la fois une mise en marche, et une mise en cause' (p. 31).[5] Ricardou follows Genette in defining narrative as 'a written or spoken discourse which undertakes to relate an event or series of events' (p. 26). He then argues that all narratives have two dimensions. The referential dimension is the imaginary universe summoned up by the words, and in which the reader is invited to believe. The literal dimension is that of the language which is used to create the illusion of reality. The relationship between these two dimensions, however, is an uneasy one. Ricardou illustrates this by discussing Flaubert's description of Salomé as she comes in to dance. In the fictional world – the referential dimension – she is visible all at once; but, in the literal dimension, language is obliged to describe her visual aspects successively. The reader can only be aware of one dimension at the expense of the other. Either we accept the illusion and ignore the successivity of the writing; or we become aware that the simultaneous is being represented successively and this awareness of artifice destroys the illusion of representation. Fiction tends in one of two directions. In one type the referential dimension is uppermost: an illusion of reality is created and the material reality of the text is effaced. Alternatively, the literal dimension predominates: the illusion of reference to a fictional reality is repeatedly shattered. The New Novel is of this second type.

The rest of the book illustrates this initial definition of the New Novel. Pride of place is given to the techniques by which the New Novel undermines the illusion of representation. In various ways the text may draw attention to itself. It may be too visibly, too 'aggressively' structured, thus creating patterns of coincidence which clash with the formlessness normally associated with real life. Or such structuring may take the form of *mises en abyme*, that is to say, secondary subjects within the novel which condense the main subject and reflect it in miniature.[6] Or the language of the text may perform a similar function: the illusion of consistent reference may be disrupted by frequent and sudden transitions from one element of fiction to another, motivated by analogies of sound or sense. Alternatively, the illusion may be broken by instability within the referential dimension of the narrative. Different versions of the same incident may contradict one another; or levels of reality may be mixed: the apparently real may turn out to be the description of a film, a statue, a photograph, a painting, a postcard – or vice versa. Finally, the illusion may be threatened when the reality of passing time is too much suspended: the narrative may get bogged down in lengthy descriptions, in attempts to represent either a single incident by repeated approximations or several incidents simultaneously.

From this brief summary of Ricardou's ideas, it will already be apparent how much the previous chapter on *La Route des Flandres* owes to his example. My reading of that novel emphasised a conflict between what Ricardou calls referential and literal dimensions. By drawing attention to itself, language breaks the illusion of a stable fictional world. It does this in many of the ways which Ricardou defined, above all by playing with resemblances and variants, whether of story – de Reixach, his ancestor, the peasants – or of imagery: metaphors of eating, for example, or puns on the word 'tête'. What makes the foregoing analysis of *La Route des Flandres* not totally Ricardolian may at first be taken for mainly a matter of flavour or tone. The conflict of dimensions was presented there as a failed quest for knowledge and identity. The breaking of the illusion of representation was seen as the failure of language to name the real, experienced as tragic loss – although the loss is in part offset even in *La Route des Flandres* by a paradoxical gain: wonder at the power of language which brings the unexpected together and, while conscious of its own illusion, imitates the constantly changing shapes of life. This reading has an affective colouring which is alien to Ricardou.

There is more at stake, however, than this colouring. The subject-matter of the chapter was itself alien to Ricardou. Although I was concerned in large part with Simon's suspicion of the conventional characteristics of representation in fiction – character and plot – I traced that suspicion predominantly to an aesthetic of realism, a concern with memory and perception. That aesthetic and these concerns, shared to varying extents by other New Novelists in the 1950s, attached Simon to a form of literary and philosophical enquiry which reached France from Germany in the 1930s. Phenomenology, which Sartre and others studied in Husserl and Heidegger, finds the source of all knowledge in experience. It affirms that the perceiving subject and the external world cannot be separated: consciousness is always consciousness of something. In a study of the New Novel much influenced by phenomenology, John Sturrock summed up as follows some aspects of phenomenology particularly relevant to the New Novelists:

> Phenomenology sets out to eliminate the old and damaging dichotomy between realism and idealism, first by reducing the world transcendental to consciousness to its appearance in consciousness (speculation as to the unknowable essence of things being wholly vain) and then by denying the self any separate existence from the succession of mental experiences that constitute it. Perception, retention and memory are portrayed as relations; no longer is there me, on the one hand, and that tree, on the other, there is only me seeing, retaining, or remembering that tree.[7]

The tree which Sturrock here has in mind is not just philosophy's traditional example but probably also the tree-root of Sartre's novel *La Nausée*. Contemplating it, Rollebon loses all sense of his separateness and becomes simply consciousness of what is. Thus, towards the end of the novel, Rollebon reaches the point from which Simon's characters of the 1950s start. He has had to learn that the past cannot be forced into logically ordered patterns. His 'adventures' have disintegrated into confused and fragmentary memories, not least because perceptions themselves are fragmentary and confused. Now he stands as helpless and dismayed as Simon's characters before the swarming confusion of the world presented to him by his senses, and as captivated by its isolated details. His sense of identity threatens to dissolve. Phenomenology, then, points to a measure of literary affinity between the Sartre of the 1930s and the Simon of the 1950s. In philosophy, the phenomenological tradition continued into the post-war years. The most elaborate attempt to describe the nature of pre-reflexive consciousness when rational and scientific grids of interpretation have been bracketed out was made by Sartre's fellow-founder of *Les Temps modernes*, Maurice Merleau-Ponty, in *Phénoménologie de la perception* (1945). In the later 1950s

Merleau-Ponty became interested in Simon's novels, in which he found literary confirmation and elaboration of his own ideas. The previous chapter of this book, although working outwards from the novels themselves, runs parallel to much criticism of Simon which has taken phenomenology, and more particularly Merleau-Ponty, as an explicit frame of reference.

The parallels lie not just in the matter of the last chapter but also in the way in which it is treated. The phenomenological study of perception seeks to establish what is common to all perceivers, but it also acknowledges that 'me seeing, retaining or remembering that tree' is different from you doing the same. Hence, although phenomenological criticism pays close attention to technique and language, it also reports on what it perceives to be the philosophical implications of the writer's formal choices, especially since the way in which we perceive the world constitutes our particular vision of the world. Indeed, in a well-known article on *Le Palace*, the philosopher Michel Deguy remarked that 'Weltanschauung and the description of appearances are one and the same thing, as the ambiguity of the common expression "a way of looking at things" indicates.'[8] Merleau-Ponty himself noted how Simon's use of the present participle and intermediate forms between 'il' and 'je' weaken the authority of the self and give language a new autonomous status.[9] Much of Sturrock's analysis of Simon deals with the 'negative' foundations of the philosophy which he abstracts from the novels. Deguy found Simon's world fragmented and chaotic, a world of representations emerging from nothing and vanishing into nothing. In the 1990s Jean Duffy returned to Merleau-Ponty to explore more systematically the common ground between his theories and Simon's fiction. She too brings out how the pre-reflexive perceptual overloading characteristic of experience in some Simon novels dissolves the self, challenges conventional anthropocentrism and undermines human pretensions to free will and self-determination.[10] All these studies, like the previous chapter, imply a perceiving, thinking, feeling subject from which the text originates. They differ in two ways from Ricardou.

First, Ricardou's attention to form is more radical and more exclusive. Ricardou shares the widespread ambition of the Structuralist critics of the 1960s to understand and set out general laws which underlie the production of individual works. He also shares their confidence that such an ambition can be achieved by applying to the study of narrative some of the key concepts of Saussurian linguistics, in particular that meaning lies in networks of relationships. Like all the critics of the period he uses Saussure's terminology of sign, signifier, signified, and referent, modifying their sense to suit his purposes. There are equally aspects of Ricardou's system inherited from the linguistic base of Czech Formalism. Jakobson linked the syntagmatic axis of language – the order of words in the sentence – with the figure of metonymy and hence the linear succession of narrative prose; the paradigmatic axis – the selection of an appropriate word for each space in the sentence – he associated with metaphor, with the use of analogy to bring together like with like, and hence with poetry and the modern novel.[11] Ricardou assimilates and develops these concepts in his own way. The referential dimension is that of conventional linear prose; the literal dimension works above all by analogies and is the mark of the truly modern novel.

What distinguishes Ricardou's work from such pragmatic Structuralists as Genette or Todorov, however, is that his narratology has an explicit ideological base.[12] This leads to the second major difference between Ricardou's criticism and phenomenological criticism of which the previous chapter is broadly an example. If the previous chapter reads so differently from the summary of Ricardou's views with which this chapter began, it is because different assumptions are being made. Briefly speaking, I interpreted Simon's development culminating in *La Route des Flandres* as the expression of a crisis in Humanism. Despairing of reason and progress, Simon finds no certainty to fill the void. He discovers that language cannot be used as an instrument for getting at truth. In one sense this is the point at which Ricardou's analysis starts. He condemns as a Humanist delusion the idea that language is an instrument wielded by an author who is proprietor and master of his text. The idea that the author is someone with something to say and that he uses language to say it downgrades or even effaces the work which produces the text. It becomes merely the representation or expression of something given in advance. By thus distracting attention from the work of production, these 'dogmas of representation and expression' disguise assumptions and preconceptions of the dominant ideology which serves the interests of the ruling class.[13]

Ricardou only passingly debates what these assumptions are or how they serve the ruling class. Nor does he analyse in depth the notion of 'dominant ideology'.[14] In one sense he was not called upon to do so: that argument was being conducted for him in the 1960s by others, by the Marxist philosopher, Louis Althusser, and by the *Tel Quel* group of writers, critics and theorists to which, until 1967, Ricardou formally belonged.[15] It remains true however that Ricardou largely assumes the truth of these ideas, and they strongly – all the more strongly? – determine his approach to fiction. Thus his diagnosis of two types of novel in *Le Nouveau Roman* is in no sense even-handed. Ricardou believes that subversion of accepted artistic forms is a political act and that the transformation of forms, by altering the mental universe in which we move, contributes to transforming society. To denounce the illusion of representation, for example, is a political and moral imperative. The New Novel is morally superior to novels which fail to do this; and the best New novels are those in which the literal dimension most dominates, that is to say novels which do not represent or express some pre-existing sense and which most advertise the processes of their production, in short, which lack all traces of an originating subject. Thus Ricardou seeks to combine the materialism of Marxism – words are the writer's material which must not be overlooked or mythified – with Marxism's humanistic commitment to enlightenment and progress.

Ricardou's passionate commitment to this cause is what lends spice to the debates of the 1970s at Cerisy. In the discussions which follow each paper, faithfully recorded in the conference proceedings, the sparks fly. In 1971 at the New Novel colloquium and in 1975, face to face with Robbe-Grillet, Ricardou did not have it all his own way.[16] The Simon gathering was rather more one-sided. Ricardou and his thought-police sniffed out and denounced traces of the dominant ideology in the discourse of successive critics.[17] But the questions fleetingly or timidly raised at that conference have not gone away. Chief among these: can all traces of the originating subject be excised?

To this question Ricardou, in his more theoretical works, gives a slightly ambivalent answer. At the beginning of *Le Nouveau Roman* he argues that all fiction has both referential and literal dimensions; it is merely their relative importance which differs from novel to novel. Similarly, the referential dimension is acknowledged in the first part of his definition of the New Novel: the narrative is 'put in motion'. In what follows, however, he uses a number of manoeuvres to suppress representation and expression. One of these is to enlist some enemy troops into his own army. The idea of 'putting in motion' implies that there must be some techniques whose function it is to create the illusion of representation. In fact Ricardou discusses a number of these – variants of the same scene, description – while arguing that, apart from a little backsliding, their function is to contest representation. Second, he plays on the word representation. When arguing that New Novels unmask the illusion of representation, he is using the word in a particular and limited sense, meaning a coherent fiction. But even if a fiction is incoherent, if elements of story are self-contradictory, so confusing that they cannot even be attributed to the mind of a confused protagonist, does this necessarily negate all sense of expression or representation? In fact, Ricardou's schematic division of narrative into two dimensions is accompanied by an equally schematic reduction of its effects: the two dimensions are in conflict, therefore fiction either creates an effect of representation or breaks it. This analysis takes no account of the effects either of particular fragmentary representations or of the whole range of effects which may be created when these fragments are juxtaposed and interwoven with one another.

Yet another problem concerns the source of such effects. Ricardou by no means excludes relationships between texts. At the Simon conference at Cerisy he introduced the notions of 'general intertextuality' and 'restricted intertextuality' to describe relations between texts by different authors and those between texts by the same author. But his interest in intertextuality is limited almost exclusively to how one text, or fragment of text, may be used to generate another and be subsequently transformed in the production of the new text. Ricardou does not recognise that the space of language is one of multiple dimensions and that words, motifs, images and themes carry all kinds of resonances with them from text to text. To put this another way, more literally, Ricardou reduces the role of the reader, because it is of course he or she who carries the resonances and brings to Simon the codes which create varieties of effect. It would be no more than a slight caricature to describe Ricardou's ideal reader of New Novels as a person obsessed by a single effect: the breaking of the referential illusion; and by a single urge: to classify techniques for doing this into the maximum possible number of categories and sub-categories. Given this definition, Ricardou himself is probably the ideal reader of New Novels, although some of his followers have run him pretty close.

One can better gauge Ricardou's strengths and weaknesses as a commentator on Simon, and look at some of his other arguments, by taking the example of his analysis of *La Bataille de Pharsale*. His key article, 'la bataille de la phrase'(1970), is typical of structuralist narratology in that it begins by enunciating general rules of narrative construction. Here Ricardou highlights an argument that takes second place in *Le Nouveau Roman*, that in certain novels the illusion of representation is broken bacause the literal apparently generates the referential. In *Le Nouveau Roman* he gives examples of structures given in advance which determine how stories must unfold: Butor's schema for his novel *La Modification*, or Robbe-Grillet's plan to invert the structure of the Oedipus myth in *Les Gommes*. But in the Simon article, as in other essays, the principal means of such generation are presented as the various ways in which language can be taken literally. The basic building-blocks of textual production are single words and metaphors. Words play a generating role through their multiple meanings and the associations suggested by their sounds. Metaphors become 'structural' when their vehicle – the image introduced as a comparison – is not simply used and then discarded but carries the story off in a new direction.

La Bataille de Pharsale begins with a concentrated cluster of images:

> Jaune et puis noir temps d'un battement de paupières et puis jaune de nouveau: ailes déployées forme d'arbalète rapide entre le soleil et l'oeil ténèbres un instant sur le visage comme un velours une main un instant ténèbres puis lumière ou plutôt remémoration (avertissement?) rappel des ténèbres jaillissant de bas en haut à une foudroyante rapidité palpables c'est-à-dire successivement le menton, la bouche, le nez, le front pouvant les sentir et même olfactivement leur odeur moisie de caveau de tombeau comme une poignée de terre noire entendant en même temps le bruit de soie déchirée l'air froissé ou peut-être pas entendu perçu rien qu'imaginé oiseau flèche fustigeant fouettant déjà disparue l'empennage vibrant les traits mortels s'entrecroisant dessinant une voûte chuintante comme dans ce tableau vu où? combat naval entre Vénitiens et Génois sur une mer bleu-noir crêtelée épineuse et d'une galère à l'autre l'arche empennée bourdonnante dans le ciel obscur l'un d'eux pénétrant dans sa bouche ouverte au moment où il s'élançait en avant l'épée levée entraînant ses soldats le transperçant clouant le cri au fond de sa gorge
>
> Obscure colombe auréolée de safran

> Sur le vitrail au contraire blanche les ailes déployées suspendue au centre d'un triangle entourée de rayons d'or divergeants. Ame du Juste s'envolant. D'autres fois un oeil au milieu. Dans un triangle équilateral les hauteurs, les bissectrices et les médianes se coupent en un même point. Trinité, et elle fécondée par le Saint-Esprit. Vase d'ivoire, Tour de silence, Rose de Canaan, Machin de Salomon. Ou peint au fond comme dans la vitrine de ce marchand de faïences, écarquillé. Qui peut bien acheter des trucs pareils? Vase nocturne pour recueillir. Accroupissements. Devinette: qu'est-ce qui est fendu, ovale, humide et entouré de poils? Alors oeil pour oeil comme on dit dent pour dent, ou face à face. L'un regardant l'autre. Jaillissant dru dans un chuintement liquide, comme un cheval. Ou plutôt jument.
>
> Disparu au-dessus des toits. (BP, 9–10)[18]

Ricardou points out that these paragraphs need not be considered the novel's true beginning: they are preceded both by the title and by the epigraph, taken from Valéry's poem *Le cimetière marin.*

I

ACHILLE IMMOBILE A GRANDS PAS

> Zénon! Cruel Zénon! Zénon d'Elée!
> M'as-tu percé de cette flèche ailée
> Qui vibre, vole, et qui ne vole pas!
> Le son m'enfante et la flèche me tue!
> Ah! le soleil. . . . Quelle ombre de tortue
> Pour l'âme, Achille immobile à grands pas![19]
>
> Paul Valéry

Ricardou argues that title and epigraph together generate the novel. He shows first that many of Valéry's words are taken up in the opening paragraphs: 'vibre' appears directly in 'l'empennage vibrant', indirectly in 'le bruit de soie déchirée l'air froissée', 'une voûte chuintante', 'l'arche empennée bourdonnante', and 'un chuintement liquide'; 'vole' in 'Ame de juste s'envolant', and also in 'jaillissant de bas en haut'; and so on with 'ne vole pas', 'tue', 'soleil', 'ombre', 'âme' and 'Achille immobile à grands pas'. These expressions, and those generated from them in the first paragraphs, then become 'directives' which dictate what follows. For example: 'le jaune' – derived from 'soleil' – 'devient l'une des exigences à laquelle doit se soumettre *la Bataille de Pharsale*': yellow as a colour, as a symbol of cuckoldry and as a series of letters open to transformation by anagram. Other words of the epigraph are exploited after the first paragraphs: 'Zénon, Achille déterminent l'exigence de la Grèce, satisfaite par Pharsale: bataille et sèjour de O.' The title itself of course also bears responsibility for both these elements, as it does for others by metaphorical association:

> [L]e générateur conflit, par exemple, produit certes Pharsale et autres batailles (de celles du Moyen Age à celle de *la Route des Flandres*, en passant par Kynos Kephelia), mais aussi match de football (près de Pharsale, justement), et manifestation de rue, lutte d'enfants et coît (notamment par l'assimilation traditionnelle de la flèche empennée et du pénis. . . .).[20]

Ricardou goes on to discuss the transitions from one element of story to another once the play of generating words and phrases is under way. He contrasts the conventional linear narrative with *la Bataille de Pharsale*. The linear narrative may consist of more than one series of events, as in Saint Exupéry's *Vol de nuit*, but if it does, these remain distinguishable from one another and coherent with one another. They fulfil the reader's expectation that things will happen to modify definitively the initial situation. Such change is limited and reassuring. *La Bataille de Pharsale* does not work like this. It gives precedence to the principle of analogy. The first paragraphs provide evidence for this. A series of associations carry the text from pigeon, to crossbow, to naval battle, to dove, to stained-glass window, and finally, through similarities of shape and the Virgin Mary, to female sexuality and the splash of a urinating mare. Subsequent transitions are worked by all kinds of analogy: puns (perfect and imperfect), similarities and contrasts of sense. What is more, analogies progressively disrupt any sense of linear progression. The same elements recur in varied forms and in new permutations. So the novel does not advance but circles on itself. The final page of the novel describes a man at a desk before a page of blank paper who writes in conclusion the novel's opening words, 'Jaune et puis noir temps d'un battement de paupière et puis jaune de nouveau' (BP, 271). Ricardou comments: 'celui qui prend la plume ne saurait être un écrivain qui s'apprête à représenter ce qu'il voit ou exprimer ce qu'il sent. O dépourvu d'identité par le travail du texte, c'est, pris dans la trame et produit de son produit, un scripteur.'[21] Thus *La Bataille de Pharsale* demonstrates the productive power of language and, concomitantly, the complete effacement of the originating subject.

Ricardou's view of *la Bataille de Pharsale* is convincing as far as it goes. One may question whether the epigraph serves as such an all-powerful generator. Some elements of it – 'tortue', the name 'Paul Valéry' – seem to get lost. Some key themes of the first paragraphs – Christianity and sexuality – are barely traceable to the epigraph. Given first, the infinite analogical resources of language, second, the length of the novel and, third, Ricardou's breathtaking ingenuity, it is not surprising that he succeeds in finding so many overt and covert links. Nevertheless he undisputably brings out dominant patterns and relationships: the key place of 'jaune' or the metaphorical importance of battle. Similarly, although his hypothesis of the linear novel is a crude caricature, his insistence on analogy, his account of the transitions between narrative sections and his analysis of the novel's structure are all much to the point. The novel is in three parts. The title of Part Two – 'Lexique' – signals repetition and permutation as key features of the text. Here Simon treats in separate sections some of the main motifs and themes of the first part: 'Bataille', 'César', 'Conversation', 'Guerrier', 'Machine', 'Voyage', 'O'. The third Part recapitulates the same material in new permutations and more condensed form: 'on doit se figurer l'ensemble du système comme un mobile se déformant autour de quelques rares points fixes' (BP, 186).[22]

What Ricardou does is to bring to the surface a subtext of verbal relationships which might have been missed by any reader searching in vain for a coherent story or plot. He makes *La Bataille de Pharsale* intelligible as an adventure in words. To put it another way, Ricardou's article has the merit of an X-ray: it reveals the skeleton of the text and shows how the bones are fitted together. What it lacks is flesh and blood. It gives no sense of the texture, the feel, the relief of the novel. It also overplays metaphor at the expense of metonymy in the widest sense and undervalues the referential dimension.

As many commentators have noted, *La Bataille de Pharsale* is a novel of transition in Simon's work and the transition can be documented within the novel itself. Between its First and Third Parts it moves towards a new impersonality.

The First and Second Parts of the novel may broadly be seen as a recapitulation of previous novels. In style and substance the first paragraph recalls *La Route des Flandres* or, even more, since it deals with the flight of a pigeon, *Le Palace*. It moves from perception through memory and back to perception. In attempting to capture precisely the sight, feel, smell and sound of the passing bird, the sentence creeps forward through corrections and elaborations, dogged by doubt: 'peut-être pas entendu perçu rien qu'imaginé'. The predominant parts of speech, as in Simon's work from *Le Vent* onwards, are past and present participles. This introduction having created some sense of an observer, it is the function of the first two parts of the novel, like the first two parts of *La Route des Flandres*, both to elaborate and to erode that sense of a containing consciousness. In Part One the text mingles various levels of past time which are almost attributable to a single consciousness: memories of childhood, of the débâcle of 1940, of a journey in Greece, of scenes outside an apartment block and at the door of an apartment. From *Histoire*, however, the novel inherits an ambivalent freedom: these elements are juxtaposed in paragraphs which lack explicit links in memory; and the text hesitates between the first and the third persons. Who is the jealous lover who apparently had an affair with a painter's model? The first-person narrator or Oncle Charles? This overlap between two characters retrospectively casts doubt on the referential status of the first paragraph: perceptions and memories float between different possible observers.

A similar loss of certainty affects each element of the story. Each is in its own way an attempt to find knowledge which fails. The young man in love wants to know what is happening behind the closed door; the child to find the meaning behind the opaque materiality of the Latin words; the tourist in Greece to get to the reality presumed to lie behind these words: the site of the Battle of Pharsalus. None of these efforts is successful. The door does not open; the words of the Latin text open only on to other words, since written accounts of Caesar's battle are contradictory and confused; and the site cannot be found: which site, which of the various battles at Pharsalus? So words, for all their material presence, fade into thin air: 'Rien d'autre que quelques mots quelques signes sans consistance matérielle comme tracés sur de l'air assemblés conservés recopiés traversant les couches incolores du temps des siècles à une vitesse foudroyante remontant des profondeurs et venant crever à la surface comme des bulles vides' (BP,91).[23]

Part Two reinforces this sense of failure. Its Proustian epigraph, like the anouncement of Georges's visit to Corinne in *La Route des Flandres*, suggests a renewed attempt to get behind words, to treat them as signs bearing essential truths:

> *Je fixais avec attention devant mon esprit quelque image qui m'avait forcé à la regarder, un nuage, un triangle, un clocher, une fleur, un caillou, en sentant qu'il y avait peut-être sous ses signes quelque chose de tout autre que je devais tâcher de découvrir, une pensée qu'ils traduisaient à la façon de ces caractères hiéroglyphiques qu'on croirait représenter seulement des objets matériels.* (BP, 99)[24]

But what follows suggests that Part Two should rather be read in conjunction with the earlier description of hoofprints, the signs left by horses: 'la plage piétinée n'était plus qu'un enchevêtrement confus d'empreintes de sabots se superposant se détruisant les unes les autres' (BP, 34).[25] For the formal separation of individual themes and motifs in this section turns out to be illusory. Stylistically, thematically or referentially, they overlap and interweave both with Part One and with one another. 'César' is not just Caesar and the site of the Battle of Pharsalus, but also, by association, childhood, Christianity, female sexuality, decay, death, Flanders in 1940; varieties of shape: oval and rectangle; and varieties of representation, on locket, coin and banknote.

Announced in the final section of Part Two – 'Repartir, reprendre à zéro' – , Part Three, from its beginning, eschews the emotive and the subjective:

> De l'autre côté de la vitre, des prés, des bois, des collines, dérivent lentement. La vue est parfois obstruée ou déchiquetée par le passage rapide de talus ou d'arbes qui bordent la voie. La tête de l'un des Espagnols, penché en avant, se découpe sur le fond lumineux et changeant de la campagne emporté dans un mouvement horizontal. Le profil à contre-jour est d'un dessin aigu, le nez en bec d'aigle. Les cheveux sont cosmétiqués et lissés en arrière. Le regard est dirigé sur le gros Espagnol, assis sur la banquette en face, vêtu d'une chemise marron et qui parle sans discontinuer, les deux autres – le maigre de profil et le chauve – se contentant d'acquiescer de la tête chaque fois que le gros termine une de ses phrases par le mot No prononcé d'une façon interrogative. Toutefois cette interrogation qui se répète à intervalles rapprochés (toutes les deux ou trois phrases) semble être un tic de langage et ne pas appeler autre chose que les imperceptibles hochements de tête qu'elle provoque. (BP, 189)[26]

This scene, no less than that which begins the novel, implies an observer whose presence is signalled in the reference to disrupted vision and in the interpretation of the Spaniard's 'tic de langage'. If however one discounts the whiff of comic irony – two compliant Spaniards held in thrall by a bulky Ancient Mariner – what distinguishes the writing of this scene is the precise neutrality of tone and the conventionally ordered syntax. No disruptive ramifications suggest an anguished search for the right word. Reconstruction of the past has ceased to be an issue: the dominant present tense eliminates any potential sense of distance in time between the observer and the observed. In the course of Part Three the point of view shifts several times from one observer to another: from traveller to jealous lover to woman to writer. Each observer is designated by the simple letter 'O'. When the pont of view shifts, the tone remains unchanged. Simon, it appears, is intent on excising from his work all traces of the personal.

It may at first sight seem curious that Ricardou should not acknowledge this progression within *La Bataille de Pharsale* since the novel so clearly moves in a direction sympathetic to his views. But the explanation is not far to seek: to acknowledge this transition accomplished within the novel would be to concede to it in the first instance a referential dimension which Ricardou seeks to deny. He similarly fails to acknowledge the continuing importance of the apparent referentiality of description.

In the early 1960s Ricardou wrote a number of essays about what he then called the 'creative'[27] power of description as illustrated variously by his own works and by those of Ollier and Robbe-Grillet. The principle of analogy was already central to his argument but applied to the qualities of the object described rather than to individual words:

> Lorsque, par l'exercise d'une précise description, le romancier met en place (et nous l'avons vu, isole) l'une des qualités d'un objet (un triangle par exemple) cette forme *libre* suscite latéralement toute une série d'objets aptes à l'incarner (triangulaires). Le même phénomène se produit avec chacune des qualités de l'objet de base et entraîne des séries correspondantes. [. . .] Plusieurs fois offerts à l'écrivain, ces objets finissent par s'imposer. [. . .] Par des relais et chaînes contrôlés. [. . .] qui s'accomplissent à chaque instant et à tous les niveaux, la matière romanesque se trouve entièrement *inventée* par l'exercise de la description.[28]

During the years of their friendship, Simon spoke more frequently and more enthusiastically about this than about any other aspect of Ricardou's work.

> J'ai trouvé dans une étude de Ricardou à peu prés cette idée: lorsque nous percevons un objet, le monde entier se trouve nié, annulé. Ce verre, quand je le regarde, me prive de tout ce que je ne vois pas. Si j'entreprends de décrire ce verre, – qui me cache tout quand je le regarde, – je dois énumerer des qualités qui sont communes à ce verre et à beaucoup d'autres objets. Ce verre se trouve dès lors relié au monde qui se reconstitue dans le langage.[29]

Like Ricardou, Simon emphasises the common qualities which lead from one description to another. In recalling Ricardou's argument, however, he gives it a different twist. Ricardou emphasises the complex play of transformations; Simon highlights the outward movement whereby the pursuit of analogies leads to a reconstitution of the world in language.

The section entitled 'Machine' in Part Two of *La Bataille de Pharsale* offers a paradigm of Simon's practice and of the complexity of effects which are achieved. The opening of this seven-page, five-paragraph section is deceptively flat:

> Partant de l'essieu qui joint les deux grandes roues de fer, une chaîne (semblable aux chaînes des bicyclettes mais en plus gros) se dirige d'abord à l'horizontale vers la droite où elle contourne une roue dentelée au plateau évidé, comme celui d'un pédalier, après quoi elle remonte vers la gauche, s'infléchissant au passage sous une poulie, reprenant ensuite son ascension jusqu'à une seconde roue dentelée, plus petite que la précédente, puis elle redescend, contourne trois plateaux de différentes grandeurs et repart vers la droite rejoindre l'essieu moteur commandant aussi de façon synchrone, par un jeu de pignons biseautés, une autre chaîne, celle-là beaucoup plus robuste que la première, et qui disparaît en s'élevant obliquement à l'intérieur de la machine dans un plan parallèle à l'essieu et par conséquent perpendiculaire à celui dans lequel tournent les roues et se meut la première chaîne. (BP, 147–8)[30]

The meticulous complexity of this description prompts, one might even say, challenges the reader to visualise the machine: it fosters the illusion of representation. At the same time, it is oriented, literary, selectively ordered. The emphasis, partly on relative size, lies even more from the beginning on relationship and purpose: the axle links the wheels, and each of the following verbs personify and give direction to the static chains: 'se dirige, contourne, remonte, disparaît, tournent'. This emphasis on purpose foreshadows an outward movement in subsequent paragraphs. The conclusion drawn in paragraph two that the machine is incomplete leads to speculation about how it worked and how it got there. This in turn culminates in a burlesque vision of history as a tidal wave of 'mules, motor cars, sales representatives, contracts of sale and reaper–binders'. Receding, it had left behind

> les représentants de commerce subsistant encore quelque temps, classant leurs bons de commande et les traites signées sur la table d'un café de village, vidant avec une grimace un dernier verre de vin du pays, laissant au patron en guise de souvenir artistique quelque planche-réclame aux couleurs pimpantes ornées de machines et de jolies filles, et repartant enfin dans leur empyrée de gratte-ciel aux teintes pastels, de cancers de briques, d'usines et d'aéroports géants.(BP, 152)[31]

History peters out in contemporary, urban, mechanised society. This commentary, in style, tone and sentiment, binds the reaper (and the reader) to previous Simon novels, to *Le Vent* and *Histoire*. And not just to Simon, since the pretext for it is this meditation on the sound of a name:

> MAC CORMICK [. . .] nom que sa répétition, sa fréquence sur ce genre de machine, a vidé de toute résonance écossaise [. . .] et devenu [. . .] avec sa consonance sèche, son cliquetis métallique, comme le patronyme générique de tout ce qui, à la surface du monde, rampe dans un grincement de chaîne et de fer entrechoqué. (BP, 151)[32]

The reaper–binder has unfolded like a Japanese paper flower, and reveals itself, through idea, alliteration and sentence rhythm, to be a homage to Proust.

Ricardou deals with relatively few of the literary references in *La Bataille de Pharsale* and his concern is above all with Simon's exploitation of their properties so that, wrenched out of their old contexts, they become exclusively integrated into the new. In fact, *La Bataille de Pharsale* abounds with literary and cultural allusions, to the Bible and Catholic liturgy, to classical myth and history, to the visual arts – from the Italian and German Renaissance to twentieth-century comic strip –, and to the written word: Plutarch, Caesar and Apuleus; Elie Faure, the art historian; Valéry and Proust; a popular history of the First World War; publicity slogans and newspaper headlines. Described, quoted or mentioned, all these references are representations of representations. They bear Simon's familiar mark of failure. When the narrator searches in vain for a quotation on a left-hand page, he is looking for a source which, in turn, refers only to a more distant, inaccessible source: the experience of the writer which may or may not have resembled his. There are no essences, no final points of origin. Yet here the emphasis lies not on failure but on adventure. There is no ultimate knowledge, but a wealth of cultural representations. To write is to join the dance, to reach out and make contact with reality which is only known in language and which is language.

Simon dances; Ricardou prefers in the main to watch from the sidelines, although he plots some of the figures of the dance, in detail; other critics have been less reluctant to join in. For example, Ricardou endows the reaper–binder with limited symbolic significance; he sees its dual function as a metaphor for the two processes he studies, generation and transition.[33] David Carroll makes wider claims for it. He attacks Ricardou for presenting fiction as a perfectly functioning machine, transforming 'elements which might seem initially to exist outside it to make them function according *to its rules*'.[34] He remarks that through the wear and tear of production the binder has broken down; it thus symbolises that no text can ever constitute a perfect closure. Other critics have seen the novel primarily as a critical commentary on past fiction. What's in a name? To construct from the sound and shape of a name fantasies of its meaning is *par excellence* a Proustian technique. For those who put Simon's many Proustian references and quotations at the centre of *La Bataille de Pharsale*, the binder – 'inutilisable', 'paralysé', 'anachronique' (BP, 151–2) – becomes symbolic of the Proustian novel to which Simon is indebted and which he is now rejecting.[35]

The dance need not end there. One could read a more private significance in the machine. The decay of the machine represents not just a critique of Proust's aesthetic – suffering transcended and the past recaptured through art – but a self-critique. The last part of Section Two of the novel consists of critical reflections on the limitations of a single point of view and the impossibility of ever pinning down the experience of a moment, given the continuous flux of time and space. It is precisely this attempt to reconstruct the past from a single point of view, Simon's practice from *Le Vent* to *Histoire*, which is abandoned in Part Three of *La Bataille de Pharsale*. Instead, the frequent shifts in point of view and levels of representation illustrate the fragmentary re-use of some previous techniques, just as the farm-workers salvage spare-parts from the decaying machine.

Simon's aesthetic allows for no definitive metamorphosis: the world is not transcended, as in Proust, nor does it run down and stop. Decay is always also re-creation. And although the spiral tends, as in Beckett, to be a downward one, the re-creation has a sensuous wealth which contrasts with Beckett's minimalism.

This is apparent, for example, from another way in which the description of the machine moves outward to make contact with the world in language: through metaphor. Compared at the beginning of the first paragraph to another mechanical construction, a bicycle, the binder is progressively, ambiguously transformed into a natural object: first its parts – the seat seems like a lily leaf, the brake-handle like a beak, the broken, unplugged cables like parasitic, metallic vegetation. Then the whole, in its imagined, reconstructed movement, becomes an insect:

> devant être capable, dans un cliquetis métallique d'élytres, d'articulations et de mandibules, d'une série d'opérations synchrones, à la façon de ces insectes capables en même temps de mouvoir de multiples pattes, de fléchir leur cou à droite et à gauche, d'ouvrir et de refermer leurs pinces, le tout s'arrêtant soudain, tout à la fois (c'est-à-dire l'insecte cessant d'avancer, de mouvoir sa tête et ses pinces) à l'approche d'un danger ou à la vue d'une proie possible et alors le silence, la terrifiante immobilité qui précède la course, la ruée. (BP, 151)[36]

This sequence of metaphors reinforces the prestige of the natural over the decadent, modern industrial. It becomes even more selectively, privately orientated in the final description of the decaying machines which strew the plains and hills:

> elles [les mécaniques démantibulées] laissent peu à peu entrevoir leurs anatomies incompréhensibles, délicates, féminines, aux connexions elles aussi délicates et compliquées. Leurs articulations autrefois huilées, aux frottements doux, sont maintenant grippées, raidies. Elles dressent vers le ciel, dans une emphatique et interminable protestation, vaguement ridicules, comme de vieilles divas, de vieilles cocottes déchues, des membres décharnées. (BP, 152–3)[37]

Derelict and abandoned, the machine is finally humanised. Its silhouette against the sky recalls de Reixach's raised sabre; a heroic, melodramatic, derisory protest against its own destruction. But the machine is also femininised. And this too is a familiar figure in Simon: woman as old crone: incomprehensibly other, repugnant yet fascinating, inviting and threatening.

The discovery of the failure of language dramatised in *La Route des Flandres* is definitive in the sense that all Simon's subsequent novels are written in consciousness of that failure. Simon abandons hope of recapturing the real and hence progressively the characteristic form of the novels in which that attempt was made: a reconstruction of the past from the point of view of a single observer. *La Bataille de Pharsale* bears traces of that old ambition. Like the novels which immediately precede it, *Le Palace* and *Histoire*, it partially re-enacts the search for the key to the enigma, for the lost centre, for completeness of knowledge, even if from the outset the search is known to be vain. But this re-enactment, though not without pain, has nothing of the anguished tone of *La Route des Flandres*. If, in one sense, language refers only and endlessly to itself, in another, it conjures up the world, since in retelling reality we recreate it. Hence all manners of exploring how language works are valid, from verbal play to effects of representation.

Ricardou's view of Simon is fundamentally flawed because he argues a priori that the literal and referential dimensions are necessarily in conflict. To make Simon truly modern, in his terms, he elides all that could be construed as referential in *La Bataille de Pharsale* and views the novel as originating and evolving exclusively as a game with words. Undoubtedly in novels from *La Bataille de Pharsale* to *Leçon de choses* Simon probes the sound and sense of words, and exploits their multiple associations with a new freedom. He gives priority to the principle of analogy and this can and often does disrupt referential continuity, the reader's sense of a coherent fiction. But Simon's art remains highly representational. It is composed of fragmentary representations. Simon's method of composition combines the referential and the literal: each sparks the other. His writing is wedded to the concrete, to what is felt, heard and above all seen. To describe – objects, actions, landscapes – is to gain access to the world of representations. This practice, increasingly dominant in the novels of the 1970s, becomes, in *Leçon de choses*, a programme:

> La description (la composition) peut se continuer (ou être complétée) à peu près indéfiniment selon la minutie apportée à son exécution, l'entraînement des métaphores proposées, l'addition d'autres objets visibles dans leur entier ou fragmentés par l'usure, le temps, un choc (soit encore qu'ils n'apparaissent qu'en partie dans le cadre de tableau), sans compter les diverses hypothèses que peut susciter le

spectacle. (LC, 10)[38]

So the description of a room, from which *Leçon de choses* begins, fans outward to the world. How it fans depends on what Simon elsewhere called 'la dynamique interne de l'écriture'.[39] Elaborating a novel from a description means for Simon exploiting the words of which that description is written. That is why description and composition are in a sense the same thing. No description in Simon is ever the innocent reflection of a given reality: all are composed, worked over. But the illusion of mimesis remains: for Simon, a point of departure, a point to which he always returns; for the reader, a constant stimulus to visualise scenes of everyday life, scenes of eroticism and jealousy, scenes of battle, contrasted landscapes and environments – urban and rural, French and American – contrasted centuries, ways of life and social classes.

Ricardou bans the referential because he wants to exorcise any possible traces of an originating subject representing the world or expressing itself in the text. To do so he identifies representation with global narrative coherence. This then gives a standard for judging Simon's novels; the best is that which most successfully denies the reader the illusion of a coherent fiction. This standard progressively forced Ricardou to revise his views of the novels. Implicitly acknowledging that there was some narrative coherence in *La Bataille de Pharsale*, he later found *Les Corps conducteurs* a more satisfying achievement, to be surpassed in its turn by *Triptyque*.[40] Analysis of the machine fragment, however, shows us that characteristic configurations suggesting traces of origin need not depend on a coherent fiction; they may be found, for example, in patterns of imagery. As Simon expressed it in 1967 when he was already working on *La Bataille de Pharsale*: 'Je crois qu'on peut aller à tout en commençant par la description d'un crayon. Le monde écrit n'est pas le monde perçu. Mais par le langage on arrive à des découvertes: on se "découvre" d'ailleurs soi-même en écrivant, dans tous les sens du mot, et c'est un risque à prendre.'[41] To write is to discover and to reveal oneself, however debatable the nature of the revelation, however dependent on the reader's subjective perception.

In 1974 at Cerisy one speaker, incautiously, found strong traces of memory and emotion in Simon's work. Ricardou immediately chided him for thinking in terms of transmission when what was at stake was a problem of transformation 'C'est cela qui est le plus difficile à penser: la transformation.' But Irène Tschinka intervened to remark: 'Ce qui n'est pas pensé ici, c'est la dialectique entre ces deux procédés.'[42] That is indeed the crux. Ricardou and those working in his wake (and all critics of Simon have worked in his wake) have done much to illuminate Simon's transformation of forms. But equally much is transmitted in Simon's work: commentaries on other texts, political and ideological preconceptions, personal and family history, psychological phantasms. How can transformation and transmission be studied together and neither be sold short? Recent criticism of Simon has begun to tackle this question; some of it will be considered in later chapters. But such criticism depends on an assumption that not all critics are prepared to make: that texts signed by Simon are in some peculiar sense Simon's texts. This is the subject of the next chapter.

3

Hierarchy and coherence: from *La Bataille de Pharsale* to *Leçon de choses*

During the 1970s and well into the 1980s, critics of Simon's works from *La Bataille de Pharsale* to *Leçon de choses* could be divided roughly into two camps. There were those who found hierarchy, coherence and unity in these novels, for whom they were 'des ensembles centrés'.[1] For others, Simon's novels of this period were increasingly discontinuous, non-hierarchised, shifting, uncentred, open, 'writeable'.[2] There is nothing surprising about this division. Much British and some American criticism of Simon had sprung from strong traditional humanistic roots nourished by the New Criticism. While drawing sustenance both from the insights of Russian Formalism and from Structuralist analyses of narrative, it had resolutely declined to swallow them whole. As a result, it presupposed, sought and found meaning in Simon's novels and various kinds of complex coherence. Progressively, however, and with increasing speed, the literary scene had been changed by the very different presuppositions of deconstructive criticism. Partly under the influence of Barthes, the fragmentary had become a value in itself. Following Bakhtin, Kristeva had taught that no text is an autonomous whole but rather a place where other texts and various codes criss-cross. With Derrida, in so far as his thought can be accommodated within traditional boundaries, the aim of criticism became to set the text against itself by teasing out inconsistencies and self-contradictions.[3] Given this variety of critical practice, it was not even surprising that some critics should appear to have a foot in both camps. But while it is possible that their assumptions were inconsistent or self-contradictory, this chapter starts from an alternative hypothesis, namely that Simon's texts justify the variety of these attempts to label them. I shall explore some of these terms and, by setting them in relation to one another, try to establish ways in which they can suitably be applied to Simon's novels from the late 1960s to the mid-1970s.

The novels of this period were seen as discontinuous, fragmentary and non-hierarchised, not least because they advertise themselves as such. *La Bataille de Pharsale* comments on its own procedures and aesthetic preferences in passages which offer opposing views of the German artists of the Renaissance:

> *on dirait que la nature est restituée pêle-mêle dans l'ordre ou plutôt l'absence d'ordre où elle se présente [. . .] tout pour l'artiste allemand est au même plan dans la nature le détail masque toujours l'ensemble leur univers n'est pas continu mais fait de fragments juxtaposés on les voit dans leurs tableaux donner autant d'importance à une hallebarde qu'à un visage humain à une pierre inerte qu'à un corps en mouvement dessiner un paysage comme une carte de géographie apporter dans la décoration d'un édifice autant de soin à une horloge à marionnette qu'à la statue de l'Espérance ou de la Foi traiter cette statue avec les mêmes procédés que cette horloge.* (BP, 174)[4]

The tone of this passage, quoted from Elie Faure's *History of Art*, clearly conveys Faure's disapproval. To his mind, these German artists lack any sense of order, of appropriate hierarchy among the elements of creation. But Faure's disapproval is violently countered by the character reading his book: 'O sort de sa poche un stylo-mine et écrit dans la marge: Incurable bêtise française' (BP, 238).[5] It would be an unsympathetic reader of *La Bataille de Pharsale* who did not take the side of Faure's critic in this dispute since the characteristics of German Renaissance painting, as Faure describes them, appear to conform so closely to those of the novel he is reading.

'Un univers fait de fragments juxtaposés'; Simon had long been fascinated by the heterogeneous. It appears as a motif in the clutter of Herzog's bedroom in *Gulliver*, the disparate contents of Marie's toffee tin in *L'Herbe* and in the undifferentiated list of events in her account book. In *Le Palace* the listing of random, unrelated events and objects is given pride of place in the titles of the symmetrical first and last chapters, 'Inventaire' and 'Le bureau des objets perdus'. Not until *La Bataille de Pharsale*, however, does any novel as a whole give the impression of being formed from juxtaposed fragments floating free of any containing consciousness. Most obvious are the fragments of narrative: a sequence concerning characters and events in a particular setting may at any moment be displaced by another sequence concerning quite different characters and events in a different setting. But there are also fragmentary quotations from a heterogeneous range of authors: Elie Faure rubs shoulders with Proust, Plutarch, a popular history of the First World War, a sequence from a comic strip. In *Les Corps conducteurs* there are fewer quotations but the same heterogeneity in the range of pictorial images from which the text originated: a photograph of a telephone kiosk, an anatomical plate, an erotic drawing, classical and modern paintings. Partly in consequence, this novel is marked by a variety, almost a jumble of registers and tones: the factual encyclopedic, the grandiose mythic, the erotic, the analytically critical, the naïve historical, the oratorical. Yet in *Triptyque* the range of registers is much narrower; and in *Leçon de choses* only the colloquial stands out as contrast and relief against a contrary, or rather complementary trend in Simon's writing. The heterogeneous has given way to the homogeneous, as demonstrated in the following passage from *Triptyque*:

> La femme est coiffée d'un chapeau de paille jaune foncé dont les larges bords sont rabattus de chaque côté de la tête par un foulard sombre passé sur la calotte et noué sous le menton. Des mèches grises en désordre s'échappent de la coiffe et retombent sur le front. Tout le bas de la figure et le menton saillent comme chez certains singes ou certains chiens. Sous la jupe flasque qui bat les mollets on aperçoit les chevilles maigres sur lesquelles tirebouchonnent des bas noirs. Les pieds sont chaussés de gros brodequins d'homme sans lacets. Les manches du caraco noir pointillé de pastilles grises sont retroussées et laissent voir les avant-bras osseux recouverts d'une peau jaunâtre. Au bout de l'un d'eux, horizontal et à angle droit par rapport à l'aplomb du corps, pend un lapin au pelage gris perle tenu par les oreilles, tantôt parfaitement immobile, tantôt agité de soubresauts et de coups de reins impuissants. Sortant de l'autre main aux doigts noueux et jaunes on peut voir par instants briller la lame d'un couteau. La fille couchée dans le foin accompagne de coups de reins le va-et-vient rythmé des fesses de l'homme dont on voit chaque fois briller le membre luisant qui disparaît ensuite jusqu'aux couilles entre les poils touffus, noirs et brillants, bouclés comme de l'astrakan. Le couple se tient dans une zone de pénombre à l'écart du cône lumineux que projette à l'entrée de l'étroit passage entre les murs de briques un réflecteur fixé au sommet d'un poteau métallique aux poutrelles entrecroisées. (T, 24–5)[6]

In one sense this is a passage of juxtaposed fragments. Context instructs the reader that within this passage there are two changes of scene, from an old woman holding a rabbit, to a couple lying in the hay in a barn, to another couple leaning against a brick wall at night in the rain of a grim Northern town. But the effect of difference is overlaid by the similarity of treatment ('traiter cette statue avec les mêmes procédés que cette horloge'), a treatment which contrasts with *La Bataille de Pharsale*, with *Les Corps conducteurs*, and even more strikingly with Simon's earlier novels. The prose of Simon's middle period, from *Le Vent* to *Histoire*, conveyed a sense of struggle, an attempt to capture, arrest and order a flood of memories and sense impressions, of words which sprang to the pen as he wrote. Here the detail of description is as precise, but it is a description without strain. The tone is uniformly unemotive. The sense of depth created by varied tenses has given way to the flatness of an unchanging present. The syntax, once precariously ordered, constantly under threat from additions and corrections, barely preserving subordination by the expedient of multiple parentheses, is now regulated and controlled in short sentences, simply constructed. In this passage there are no apparent hierarchies, of time, of intensity, of relative importance. Where English usage tends to call for possessive pronouns – her head, her face, her skirt – French syntax more naturally employs the definite article. Simon systematically exploits this feature of French usage so that aspects of dress and body are seen in and for themselves, not as human possessions or attributes. Thus everything is on the same plane here. There is neither foreground nor background; or rather, since the uninvolved, unplaced observer treats every detail with the same care, everything is successively foregrounded.

And yet, as Max Silverman has pointed out,[7] this treatment of detail is not in itself sufficient to create a non-hierarchised text. One must also take into account the form of the work as a whole and in particular how the narratives relate to one another. In *La Bataille de Pharsale* frameworks crack and split. The conventional hierarchy between what is real and what is represented vanishes when the characters in a 'real' scene are suddenly described as figures in a painting (pp. 224–5), or when the horseman on a frieze threaten to ride away (p. 259). *Triptyque* and *Leçon de choses* go further in that in these novels the narratives spill over into one another. Each is an image within the other, a poster, film, painting or calendar, which comes to life and in turn contains an image which in its turn contains an image, and so on. The effect is first to abolish the hierarchy of past, present and future: a comic *mise en abyme* of Simon's puzzled readers shows two small boys poring over the fragment of a reel of film, trying in vain to determine an appropriate chronological order (T, 174–5). When in addition, as in *Triptyque*, three narratives have equal prominence, the effect is to deprive the reader of a conventional fixed centre of interest. The text seems to have become what Simon predicted in *La Bataille de Pharsale*: 'un mobile se déformant sans cesse autour de quelques rares points fixes' (BP, 186).

Given the absence of a fixed centre, the constantly shifting perspectives, the progressive abandonment of hierarchies, it becomes tempting to apply to these novels as a whole a remark from *Le Vent*: 'nous-mêmes ballottés de droite et de gauche, comme un bouchon à la dérive, sans direction, sans vue' (V, 10).[8] That experience, one might say, has been progressively textualised. In *Le Vent* it was true of the characters; from *L'Herbe* to *Histoire*, perhaps even as far as *Les Corps conducteurs*, it could be applied to the narrators; by *Triptyque* it is the reader who is tossed on a stormy sea, with no sense of direction, no overall perspective. But can one go further than this? Do these characteristics of the text imply a more fundamental absence of order? Has the writer lost or renounced control of his language, or perhaps merely recognised that he never had it? Is he the plaything of his words?

These questions arise in the case of Simon because by the mid-1960s they were already edging towards the centre of critical debate in France. In 1968 Roland Barthes declared that the author was dead.[9] By this he meant that the author should no longer be conceived of as the source of his works. Like Proust and Valéry before him he was attacking the continuing assumptions of a crude positivistic criticism: that the writer when writing is identical with the man, that the work expresses his subjectivity and that it can therefore be explained by reference to the man. Against this, Barthes insisted that the relationship between writer and text is complex and indecipherable, partly because the act of writing influences what is being said, but more profoundly because the writer's subjectivity is itself a product of language. Here Barthes goes beyond Proust or Valéry. His argument reflects the new prestige of Saussurian linguistics which was also pervading other spheres of thought. Lacan was showing how language shapes the child's accession to the Symbolic Order, Althusser how it moulds those concepts of the relationship between subject and society which constitute ideologies. Although each of these writers left an escape route, a glimpse of possible freedom for the subject, the dominant tendency of their thought was to replace the old psychological, biological or sociological determinisms with a new linguistic determinism. The risk, perhaps even the reality, was that we do not speak language, it speaks us.[10]

Ricardou's attitude to such theorising was ambivalent. He strongly shared the view that writing could not be the expression of a pre-existing self. Consistently, and very notably in his analysis of *La Bataille de Pharsale*, he elevated language to a quasi-autonomous status. Generation took place within the work and was determined by the properties of language, its sounds, meanings and connotations. In that sense the writer was dispossessed and became, instead, the scriptor, a product of language. But that is not the whole story. In the context of the 1970s, Ricardou's writings became increasingly remarkable for how little attention he paid to the role of the reader in constructing the meaning of literary works. Whereas Robbe-Grillet emphasised textual generation as play and encouraged the reader to join in that play, Ricardou emphasised rules. His characteristic form of reading a text is to find a key, from which can be derived a set of rules responsible for the elaboration of the work: 'lire, c'est se rendre attentif à l'ordre clandestin du travail textuel'.[11] But where does the key come from and how are the rules devised? Sometimes, Ricardou left that matter totally in doubt. One means by which he avoided mention of the writer was to attribute actions and intention, metaphorically, to the work itself: '*La Bataille de Pharsale* se plaît à écrire, de son exergue, la très reconnaissable formule *Achille immobile à grands pas*', 'le texte prend aussi en compte le signifiant de l'exergue', 'les fréquent dispositifs par lesquels le texte s'engendre'.[12] Or else, the question of choice could be blurred by use of the passive: 'de la masse des possibles seront de préférence extraits ceux qui obéissent à cet impératif'.[13] But sometimes Ricardou did ascribe intention to an author: 'il faut convenir que c'est en parfaite connaissance de cause que Poe fait intervenir ici une nouvelle dimension'.[14] And what goes for Poe also goes for Ricardou himself. The key to *La Prise de Constantinople* was to be found in its opening word – 'rien' – in the words and motifs on the title-page, and in the linguistic and numerical interplay between these. These generators and their combinations, however, were the scriptor's conscious choice, although others might be found: '[C]e n'est certes pas dire que maints rapports inaperçus n'aient travaillé *à l'insu du scripteur*'[15] (my italics). Ricardou claimed to exert over his own novels an authority which his critical and theoretical works sought to extend over his readers.

Simon's view of the relationship between writer and language was equally ambivalent but less imperialistic. To much of the theory he was deeply sympathetic. His exploration of the theme of memory in novels from *Le Vent* to *La Route des Flandres* had led him to conclude that language could not capture the past: by endlessly suggesting new associations and combinations it prevented the writer representing reality or expressing what he had wanted to say. In *La Route des Flandres* that experience was perceived as a disappointment and a betrayal. But increasingly Simon came to see the fertility of language as a merit: it offered opportunities to be exploited. Thus his method of work corresponded more and more to the models proposed by Barthes and Ricardou. His novels began as unrelated fragmentary descriptions. These descriptions were progressively expanded and related to one another thanks to a dynamic power which Simon attributed to language itself:

> *Chaque mot en suscite (ou en commande) plusieurs autres, non seulement par la force des images qu'il attire à lui comme un aimant, mais parfois aussi par sa seule morphologie, de simples assonances qui, de même que les nécessités formelles de la syntaxe, du rythme et de la composition, se révèlent aussi fecondes que ses multiples significations.*[16]

A statement such as this from the preface to *Orion aveugle*, published in 1970, emphasises signifier as much as signified. It marks the high point of the convergence of Simon's views on textual generation with those of Ricardou.

Yet on the other hand Simon also never wavered from the conviction that, though language proposes, the writer disposes. He repeatedly affirmed that his novels were not Surrealist experiments in endless free association.[17] To give free rein to language was at most a stage in their construction. The writer's task was then to select and edit in accordance with a project, not preconceived, but growing and changing as his work on the text advanced:

> *chaque fois qu'à chacun des mots carrefours plusieurs perspectives, plusieurs 'figures' se présentent, avoir toujours à l'esprit, pour le choix qu'on va faire, la figure initiale avec ses quatre ou cinq propriétés dérivées et ne jamais perdre celles-ci de vue, faute de quoi [. . .] il n'y aurait pas livre, c'est-à-dire unité, et tout s'éparpillerait en une simple suite.*[18]

Simon aimed to control his writing, to make his novels unified wholes. What was the nature of the coherence he sought?

First, as the above quotation makes clear, Simon acknowledges that his kind of coherence necessarily implies hierarchy: the initial figure and its derivatives must dominate the work. In this context one may compare Simon's novels of this period with what he calls the 'academic' novel.[19] The 'academic' novel uses and encourages the reader to use familiar codes specific to the novel, for example, character, story or description. These codes have a conventional, albeit variable hierarchy. For instance, either character or story may predominate, but description will always be secondary: it will serve character or story. Simon disrupts these codes and their hierarchical relationships. Hence his novels appear non-hierarchised. There are however other, more general codes which we use in reading a novel as in interpreting other linguistic messages, for example that a title concentrates and summarises, or that what is repeated is significant. These codes Simon uses. Seen in this perspective his novels are hierarchised.

Formally, for example, the title *Triptyque* had led every commentator to discuss that novel as a combination of three stories, and to treat the circus episodes as secondary; just as *La Bataille de Pharsale*, as signifier and signified, is one of the 'points fixes' around which the text revolves and its critics have skirmished. Or again, although all words are called to be 'mots carrefours', few or relatively few are chosen. Thus *Triptyque* is written under the signs of the cross and the serpent, while in the second last section of *Leçon de choses*, 'La charge de Reichshoffen', Simon plays predominantly with the military and sexual connotations of 'tireur', 'tirer' and 'retirer', 'charge', 'chargeur' and, by implication, 'décharger'.[20] But there can be as much value in scarcity: some passages are specially prominent because they are unique, untypical, particularly dense or heightened in tone. An immaterial mass threatens to engulf the sick man in *Les Corps conducteurs* and breaks the surface realism of the text. Night falls in *Triptyque*, heavy with a lyrical, philosophical dew which distinguishes this passage from the rest of the novel. The mason's boiled egg in *Leçon de choses* concentrates and reflects the room, the world and the text.

Similarly, some themes are more equal than others. Chief among these is precisely an attack on conventional hierarchies. Simon's novels of this period are perhaps above all a series of critical commentaries on the conception of a natural hierarchy which puts man and his concerns above all else. In *La Bataille de Pharsale* fragments which begin to coalesce into a story of jealousy disperse again in a minuet of shifting points of view. In *Les Corps conducteurs* the human body itself is sectioned, divided, distributed, its parts re-formed in new shapes. Thus attention is not focused on an individual's sickness and suffering but conducted ceaselessly from one body to another. The three incidents of *Triptyque* are potentially full of human interest. They appear in all their garish drama on the cinema posters. One might sum them up as: bridegroom deserts bride on wedding-night; sex-mad nanny leaves girl to drown; baroness's son in drugs scandal. Yet in the novel itself these human dramas are no more than echoes, empty gestures. Like Orion in *Les Corps conducteurs*, 'partie intégrante du magma de terre, de feuillage, d'eau et de ciel qui l'entoure' (CC, 77),[21] Simon's characters are increasingly incorporated in the urban or more often natural environment from which they grow and into which they fade. *Leçon de choses* explores this same theme in a different way. The textbook 'leçon de choses' shows a world arranged to serve man; but as that textbook appears in Simon's *Leçon de choses*, fragmented and rewritten, its certainties are exploded: man and the world are seen as together subject to a constant process of de-formation and re-formation.

A particular aspect of this theme concerns human sexuality. In earlier novels, notably *La Route des Flandres*, the sexual act is rich in significance. It is associated both with a search for epistomological certainty and, simultaneously, with a desire for oblivion. It is conventionally treated in that sexual climax is represented by verbal climax: the pace quickens, the tone rises, images whirl and explode. Particularly in *Triptyque* and *Leçon de choses*, however, the homogeneity of Simon's style strips the sexual act almost entirely of cultural or personal, emotional significance. It becomes something purely material; and yet, though described with vivid precision, it is also lacking in sensual significance. Pornographic writing has its own conventional rhythm: a slow, teasing build-up to an inevitable climax. The sexual act in Simon's novels is subservient to a different rhythm; closely observed bodies come and go; there is no correspondence between sexual and textual climax. Human physical relationships are made to seem merely one activity among a thousand others; sex, in these novels, is nothing to get excited about.

Simon, then, questions hierarchies but, so long as he clings to coherence, cannot escape them: the figures must be limited; themes have their own hierarchy. This is the paradox and – if one is prepared to recognise the writer as conscious producer of his own text – the drama of these novels. In each of them, questioning hierarchy, Simon creates a new hierarchy, against his will. Time and again he renews the struggle to prevent hierarchies from hardening, to keep his mobile turning, to do away with the 'points fixes'. The impossible ideal would be a novel without hierarchy yet absolutely coherent. Simon's determination to keep the mobile turning is particularly evident, for example, in the autobiographical element in his novels from *L'Herbe* onwards. In a first phase, even up to *La Bataille de Pharsale*, there are some relatively fixed foundations on which the reader may construct a scaffold of knowledge somewhere alongside a mythical 'real' Claude Simon. In a very conventional way *La Route des Flandres* is the continuation of *L'Herbe*. Not merely is the same family put on stage but in particular *La Route des Flandres* explains how Georges lost his illusions: there was something more to it than failing the entry exam to the Ecole Normale Supérieure. *Le Palace* and *Histoire* probe successively further back in time. The origins of a reasonably consistent central character are traced back to the foetus described on the last page of *Histoire*. But already the ground is shifting. In *Histoire* and again in *La Bataille de Pharsale* the contours of this central character become blurred as his experience blends with that of his uncle. And when the references to familiar family names, Corinne and de Reixach, absent entirely from *Les Corps conducteurs*, reappear in *Triptyque* and *Leçon de choses*, they have an air of self-quotation, almost self-parody, a nod and a wink to the reader that these characters are drawn from other texts, not from life. The old mason of *Leçon de choses* is a specially humorous reshuffling of the cards. He recounts experiences of war blended from the Simon of *La Corde raide* and Georges of *La Route des Flandres* but in the style of Iglésia, the jockey, if one can imagine Iglésia having become garrulous in old age.

Can the same effect of motion be achieved within a single novel? Simon increasingly tackles this problem head-on. Not all words can be crossroads. But what if one multiplies the crossroads so that, as far as possible, all words, figures, themes are continuously present, explicitly or 'en surimpression', 'en filigrane'? Simon's attempt to achieve this effect, and some critics' attempts to come to terms with it, can be illustrated by *Les Corps conducteurs*. Ricardou analysed *Les Corps conducteurs* as a novel which, for all its fragmentation, is ultimately unified. Its narrative sequences are hierarchised: the rich man walking the North American city frames the second main sequence describing a writer's visit to South America to attend a conference.[22] Within that framework, a third sequence, telling the story of a group of soldiers traversing a South American forest, forms a *mise en abyme*: separable and coherent in itself, it gathers together and reflects in miniature aspects of theme and forms found in the main sequences.[23] A dominant set of themes runs through the novel: illness, the fear of surgery with overtones of sexual apprehension, and the desire for survival. This thematic unity reinforces the predominance of the framing narrative sequence.[24]

The contrary case for absence of unity and hierarchy has been most persuasively put by Michael Evans.[25] Like Ricardou, he associates unity with narrative sequences which are referentially cohesive; a novel may have several such blocks of story and character set in specific space and time but, by implication, unity requires that they should be distinct and hierarchically related. Evans acknowledges the force of Ricardou's argument that the narrative sequences in *Les Corps conducteurs* are related in this way: to use Ricardou's metaphor, *Les Corps conducteurs* may be seen as a tree trunk with spreading branches. But, Evans argues, the more closely one reads the novel, the more difficult it becomes to distinguish or hierarchise the various spatio-temporal dimensions, and the more one becomes aware of resemblances in theme, style and motif which run between the different sequences and erase their distinctiveness. In particular, Evans discerns blocks of description which are not integrated into any narrative sequence. 'Often they have several contradictory frames; sometimes they are totally unframed, floating in a narrative void.' They show 'no development internally or from one repetition to the next, which is why the referents are often in the form of paintings or photographs'.[26] The effect of such blocks is precisely to keep the text in motion. Potential frames appear and dissolve; echoes from every narrative sequence and description block resound in every other.

This microscopic reading, forceful though it is, illustrates how difficult it is for formalist criticism to discard coherence and hierarchy. Partly this is because the conventions of critical writing require it: Evans convinces because he deploys the rhetoric of unfolding argument, his points ranged in hierarchy of importance. To be convincing, he is obliged to isolate and name five narrative sequences and, albeit reluctantly, four description blocks. But Evans's argument takes him further than this. He discusses the importance of different characteristics within each block;[27] and he attributes special significance to the fourth block which he describes as 'a *mise en abyme* of the compositional nature of all the description blocks'.[28] A more exaggerated example of this same phenomenon is found in Ricardou: he illustrates what he calls the 'discoherence' of *Triptyque* in a unifying diagram which is rigidly schematic.[29] Formalist criticism identifies unity with narrative coherence; it can demonstrate very subtly how narrative coherence can be disrupted; but it then tends to substitute new forms of coherence and hierarchy based on formal considerations.[30]

Evans's argument also shows the paradoxical difficulty of the task Simon set himself in trying to achieve constant motion. By providing a wealth of detailed evidence, Evans more and more demonstrates the importance of certain key elements – words, phrases, motifs, themes, suggested frames – precisely because they continue to recur.[31] A last return to the lengthy passage quoted above from *Triptyque* can show Simon's heroic failure and paradoxical triumph. Some elements recur so often as to be almost continuously present, for example the yellows, greys and blacks which colour each scene, or the frequent references to male and female: 'la femme. . . . Les pieds chaussés de gros brodequins d'homme . . . la fille . . . l'homme . . . le couple.' Others echo passages elsewhere in the novel. The old woman's stockings cling round her ankles, just like the man's trousers a few pages before. The cone of light and the right angle between the rabbit and the old woman refer backward and forward in the text, to the circles, triangles and tangents of the schoolboy's geometry problem which frames this extract. The passage comes nearest to perpetual motion however in a cluster of motifs which combine the play of light, shapes with sexual connotations, and a suggestion of movement:

> *Sortant* de l 'autre main . . . on peut voir par instants *briller la lame d'un couteau* . . . l'homme dont on voit chaque fois *briller le membre luisant qui disparaît* ensuite jusqu'aux couilles . . . à l'écart du *cône lumineux* que *projette* à *l'entrée de l'étroit passage* . . . *un réflecteur fixé au sommet d'un poteau métallique.* (T, 25)[32]

Yet as an attempt to abolish hierarchy this passage is a failure since the rule of frequency continues to apply. Despite Simon's best efforts, not everything is present all the time; that which is most consistently present gains ever greater prominence. This then is the triumph: the passage coheres tightly along the axes formed by the various repetitions, similarities, reminiscences, comparisons and contrasts. Thus the repetition of male and female draws attention to a contrast between decaying age and vigorous youth, a contrast further refined and heightened by the similar gesture of rabbit and young couple: the death throes of the rabbit resemble the movements by which the young couple create life, the 'coups de rein' of the young woman and the 'va-et-vient rythmé des fesses de l'homme'. One can go further. The rabbit's movements are 'impuissants'; this evaluative word, the only one in the passage, comments ironically on the young couple's activity. It focuses attention on the old woman's knife which the recurring motifs place in such menacing proximity to 'le membre luisant qui disparaît jusqu'aux couilles': to truncate that phrase is to leave the last four words of it open to an alarming double meaning, an answer to the implied question: how much of the male member disappears? Turn where you will in this passage, light falls on sex and in particular on classic male fears and apprehensions. The extract coheres by virtue of its themes no less than by the recurrence of figures which summon them.

A superficial reading of some *mises en abyme* in Simon's novels from *La Bataille de Pharsale* to *Leçon de choses* might lead one to suppose that Simon's response to the crisis of language was one of renunciation: having discovered that language could not be tamed to say what he wanted, he decided to give it free rein. In fact the reverse is true. Simon resolves to meet language on its own ground. Since language produces meaning he will set out to exploit that capacity, making words resonate with words, combining their manifold connotations and associations of sound but mainly of sense. In *Triptyque* Simon describes a child colouring in a picture-book by drawing a continuous line back and forth across itself and across the outline of the images beneath. Similarly the continuous line of Simon's writing blurs familiar patterns, conventional contours, but in traversing and retraversing them creates new patterns which bind the old more closely together. Or, to change the image, his aim is to weave a densely-textured cloth. He forces each thread again and again through the figures of the warp so as to leave no visible loose ends. The scale of his ambition is extraordinary. It is an attempt both to control language, by exhausting its possibilities of combination, at least within a circumscribed field, and an attempt to control the reader. Simon's novels are 'open' works in Eco's sense of the term.[33] They break with established codes and set up their own. By various stratagems of internal reduplication they guide the reader as to how they should be read. Their ideal reader is one who is prepared to follow the adventures of their figures.

To what extent however does Simon achieve the control he seeks? Any victory over language is partial and temporary. The material is too rich; there are always loose ends, new possible combinations. To take one example: 'la description [. . .] peut se continuer (ou être complétée) à peu près indéfiniment' (LC, 10). The battle continues until the writer is exhausted.[34] It can have no other outcome. The end is always arbitrary. Over the reader, Simon's control is no more complete. Undoubtedly Simon forces readers to abandon old reading habits and adopt new ones. But if a reader accepts Simon's invitation to play, to follow the varying figures, nothing ensures that he or she will respect the limits of Simon's game: 'la figure initiale et ses quatre ou cinq propriétés dérivées'. Only connect: the habit is catching. Is that passage from *Triptyque* primarily, predominantly about the fear of castration? At some point the reading proposed above has set in concrete what in the text itself was fluid and fleeting. Does it perhaps tell more about the reader than about the text? Simon's novels of this period are eminently 'writeable' in that they encourage readings which go beyond the hierarchies established in the text.

In another sense, to raise the question of control is to move fully into one area of deconstructive criticism. What gulfs exist in these novels between rhetoric and practice? It is characteristic of all these texts to denounce the view that words represent reality. Whatever seems real quickly becomes representation, as film, painting, postcard or snapshot. Step by step the reader is forced back until compelled to acknowledge that words refer only to other words. Yet, despite this, Simon's novels remain profoundly mimetic, and not just in the sense that he uses language to describe the physical world in vivid detail. Beyond that, Simon shapes his novels to reflect what he sees to be realities. Philosophical realities: that man is 'one thing among other things rather than [. . .] a being placed above other things';[35] that we live in an exclusively material world subject to constant transformation; and also artistic realities. For Ricardou, *La*

Bataille de Pharsale was the story of its production. Jean Duffy has shown how much work, how much rewriting, it takes for Simon to create that kind of effect.[36] *La Bataille de Pharsale*, like other novels of this period, tells not *the* but *a* story of its production. That is to say it imitates the processes by which it came into being because Simon wants to tell us how his works are written. In general, Simon's novels of this time illustrate to the full one of the paradoxes which Frank Lentriccia claims to be at the heart of modernism: they seek to instruct us about reality, while denying that words can do so.[37]

Beneath this paradox lies another. As we have seen, Simon's chief theme is that there is no pre-ordained hierarchy in the natural world, man is not the king of creation. There is a way to demonstrate in writing that one thing is no more important than any other. It is to let words tumble out in random sequence. Yet in this instance Simon rejects mimesis. He opts instead for coherence and with it, inevitably, hierarchy. The reason for this choice is clear. Though conscious of the weight and shaping force of language Simon wishes to demonstrate his control of words. In 'The death of the author' Barthes conceded that the writer has 'the power [. . .] to mix writings, to counter the ones with the others, in such a way as never to rest on any one of them'.[38] Simon is, like Barthes, an arch-challenger of conventional codes. By mixing writings he attempts to demonstrate his freedom. Of course, *La Bataille de Pharsale*, *Les Corps conducteurs*, *Triptyque* and *Leçon de choses* may not ultimately be Simon's novels: there is no way of telling if we have free will. But Simon believes that they are, and they can be read as such. Like the writings of Barthes, Simon's novels represent a form of humanism, though some traditional humanists would object to this use of the word. These novels do not posit fixed truths about human nature or the nature of reality, to be discovered by the exercise of reason. But they do affirm that man creates meaning.

4

Les Géorgiques and intertextuality

In the 1980s Simon published two substantial novels, *Les Géorgiques* in 1981 and *L'Acacia* in 1989. In both, Simon offers a new synthesis of all that went before; but *Les Géorgiques*, because of the novels which immediately preceded it, was the greater surprise. *Les Géorgiques* shares with *Leçon de choses* the importance of description as a means of generating narrative, with *Triptyque* the imbrication of three separable narratives, with *Les Corps conducteurs* the transformation of visual images into text, with *La Bataille de Pharsale* the incorporation and reworking of other texts. But, at the same time – and this was the surprise – Simon reintroduces in *Les Géorgiques* what Jean Duffy has aptly called 'the ramificatory plot structure and hyperbolically delineated characters of works such as *Le Vent* (1957) and *L'Herbe* (1958)'.[1] These characters and their stories, like those of *La Route des Flandres*, *Le Palace* and *Histoire*, are set in particular historical circumstances; as in these novels, the past is reconstructed from flimsy surviving evidence. The style is mixed: sometimes clipped, objectively informative as in the novels of the 1970s; elsewhere, long sentences strain at the bonds of conventional syntax and feel their way forwards over pages. Frequently, an expansive narrative voice, not limited to factual notation of perceptions, comments reflectively or ironically on action and characters, and ponders the meaning of history.

Reviewers and critics gave to this 'work of synthesis and renewal'[2] a warm reception. Stuart Sykes attributed to 'many faithful readers of Simon' a reaction which the tenor of his article suggested he shared himself: 'something dangerously close to a sigh of relief'. *Les Géorgiques*, he remarked, displays a concern with getting away from the devastating circularity of the 'nouveau roman in its more recent manifestations'.[3] On this view, Simon had come too much under Ricardou's influence in the 1970s. Now he re-emerges: *Les Géorgiques* is a novel about history and nature, war and revolution, about human, or at least men's, experience of these and what they mean. In other words, *Les Géorgiques* seems to reopen the way to traditional interpretative criticism. Occasionally, it has been treated in this way. For most critics, however, *Les Géorgiques* remains primarily a self-reflexive work. In making visible its own processes of generation, it raises questions about representation and reference, about genre, about the writing of fiction, history or biography.[4] Above all critics have been fascinated by the aspect of *Les Géorgiques* which Simon signals in the novel's title: the presence in this work of the traces of so many others, mentioned, quoted or rewritten. But 'intertextuality', the word most commonly used to describe this presence, has in itself so many meanings that I intend first to consider these briefly before turning to *Les Géorgiques*.

Julia Kristeva coined the word intertextuality in 1967 when introducing to a French public Bakhtin's theory of the dialogic novel.[5] In the ferment of structuralist and post-structuralist ideas of the late 1960s, intertextuality came with an anti-humanist colouring. For traditional humanist criticism literary texts have psychic origins. Their themes can be traced back to the minds of their authors; their literary and other 'sources', albeit transmitted through writing, lie ultimately in the 'influence' of other minds. Intertextuality cuts the direct link between text and mind and places texts primarily in relation to other texts. Texts form a signifying system parallel to that, and ultimately elaborated from that, which Saussure had established for language itself. Thus in Kristeva's view the intertextual would displace the intersubjective: 'à la place de la notion d'intersubjectivité s'installe celle d'*intertextualité*'.[6] Kristeva did not deny the existence of a problematic subject, the author, who wrote the text: much of her later work would be concerned with pursuing the traces of this subject. But her coining, intertextuality, began the process by which an area was opened up for study from which other critics would more forcibly banish the inter-subjective.

By the same token intertextuality was anti-referential. Saussure had argued that signs do not name but draw their meaning from their relationship to other signs in a signifying network. From this, structuralists and post-structuralists drew the philosophical conclusion that there is no reality whch transcends the signifying network. In Derrida's famous phrase: 'il n'y a pas de hors-texte'. To attempt to name something in language then is paradoxically not to make it present but to signify its absence: 'la "réalité" est ce qui échappe'.[7] The real is always deferred, out of reach. Intertextuality makes that failure of reference visible. As soon as the reader becomes aware of it, it subverts the apparent reference to a world beyond language. Each text seems to point to another and then another within a system infinitely extendable but from which there is no escape.[8]

The broad aim of structuralist poetics was to discover the general rules which underlay the construction of particular texts. As applied to intertextuality, that aim became to define the ways in which one text can appear in another. This involved detailed work of classification, the coining of neologisms, and frequently the use of tables, since it was intended to document not just all known but all possible modes of the appearance of one text in another. The cake of intertextuality can be sliced in many different ways. For example, along the axis of visibility – from the most overt kind of intertextual link (the attributed quotation) to the most discreet (the allusion). This in turn can be distinguished from the axis of scale. In 1981, Gérard Genette proposed that the term intertextuality be restricted to the small-scale phenomena of quotation and allusion, and defined four more classes of what he generically called 'transtextuality'.[9] The class which concerns us most in this chapter is one which includes other instances of relationships between two particular texts. (Genette calls it 'hypertextuality'.) Within this class further axes can be discerned. Genette devotes much of *Palimpsestes* to parody and pastiche: these large-scale phenomena are conventionally placed on an axis of meaning which runs from the critical to the respectful. Another axis proceeds from the relative simplicity of transformation (transformation proceeds by elision or substitution), to the complexity of imitation (imitation implies understanding of the original). Finally, how is the old text, the intertext, introduced into the new? Is it, for example, naturalised as part of the fiction – a book read by a character, say – or, at the other extreme, apparently unassimilated, a foreign object challenging the reader to make sense of its relationship with the host text?

Much of the discussion undertaken by structuralist narratologists such as Genette has been based on assumptions not necessarily consistent with those of the more radical theorists of intertextuality.

In particular, those exploring the appearance of intertexts in specific host texts have tended to consider host texts as autonomous unified wholes, rather than mere crossroads where many texts meet. Although Laurent Jenny acknowledged the extra resonances of meaning which intertextual references introduce, he nevertheless defined intertextuality as 'le travail de transformation et d'assimilation de plusieurs textes opéré par un texte centreur qui garde le *leadership* du sens'.[10] Riffaterre, studying individual works, consistently tries to demonstrate that the intertexts fill gaps in sense: the informed reader is in a position to reconstruct the complete meaning of the work.[11] Such confidence also underlies the influential Freudian approach of Harold Bloom. He maintains the link between text and author's psyche by interpreting formal differences in intertextual practice as expressions of differing attitudes towards preceding writers, each marked by the anxiety of influence.[12] All such readings leave essentially unchanged the humanist, idealist conception of text and of the relationship between text and reader. That is: a text incarnates certain meanings which the reader attempts to discover. The more experienced and skilled the reader, the more chance that he or she will penetrate to its essential truth.

The position of Barthes has been both central and shifting. Although his 'Introduction à l'analyse structurale des récits' of 1966[13] was not significantly devoted to intertextuality, it shared the formalist objectivity characteristic of the writings of moderate structuralists, and was influential in promoting narratology. In 1970, however, in *S/Z*, he firmly shifted the debate away from the formalist analysis of texts towards the act of reading. Meaning, he affirmed, is never given or complete because it is constructed in each text afresh with every reading and the reader is never the same: 'ce moi qui s'approche du texte est déjà lui-même une pluralité d'autres textes, de codes infinis, ou plus exactement perdus (dont l'origine se perd)'.[14] Barthes thus widens the issue well beyond questions concerning the presence of one text in another, or even the reader's awareness of that presence. For Barthes, as for Kristeva and other semiologists, all meaning is a matter of signifying systems. Many signifying systems intermesh in language and more particularly in the written word. The codes and conventions of fiction overlap with those of philosophy or history. In *Mythologies* (1957) Barthes had studied the system of signs which, in popular culture and journalism, characterise a petty bourgeois reading of the world. Subsequently he tackled the codes of fashion.[15] Even in *S/Z* Barthes's emphasis is not on the mode of appearance of other texts in Balzac's *Sarrasine* but rather on conventions of reading and writing literary texts which make comprehension possible. Nevertheless, Barthes's definition of the nature of readers and reading influenced many subsequent studies of intertextuality, not least because to it he firmly attaches a value judgement. Complexity and ambiguity are to be preferred to simplicity and singleness of meaning. The 'writeable' text is superior to the 'readable' text because by its very open-endedness it stimulates desire, enchantment, sensual pleasure. By implication then, in terms of simple intertextuality, Barthes calls for the presence of many texts in one. The larger the number of intertexts, the more varied and complex their relationships with the host text, the more undecidable will be the host text's sense, and the greater will be its value.

Intertextual analyses of Simon novels and of *Les Géorgiques* have shown the same variety of conceptions of the nature and value of intertextuality as are found in more general studies of the topic.[16] Some have already been considered briefly in Chapter 2; others will be referred to later in this chapter. I do not intend to review them here. Rather, I want to examine a tendency on the part of some critics to doubt whether *Les Géorgiques* can be understood except in the context of Simon's previous novels and of the texts which he reworks.[17] How welcoming a novel is *Les Géorgiques*? Are new readers best advised to approach it via a detour through *La Route des Flandres*, *Le Palace*, *Histoire*, *La Bataille de Pharsale*, *Leçon de choses*, Virgil's *Georgics*, Orwell's *Homage to Catalonia*, Strachey's *Eminent Victorians* and Michelet's *History of the French Revolution*? In what sense, if any, is it true that *Les Géorgiques* must be read intertextually?

First, then, an intertextual reading of the prologue to the novel. The prologue is an untitled, unnumbered section seven pages long. It plays a role similar to that of its counterpart in *Leçon de choses* as a textual generator, a first sketch of subject, themes, motifs and words to be subsequently elaborated in varieties of tone and colour. But whereas the prologue to *Leçon de choses* begins as a naïve representation and progressively discovers that it is engaged with previous representations, *Les Géorgiques* begins with the description and analysis of a supposed eighteenth-century sketch, part drawn, part painted. Simon's tone is sober; he refrains from explicit value judgements; he does not subsequently relate the prologue directly to characters and incidents in the novel. This encourages readers to fill out its meaning in their own ways. My choice is to see it as doubly intertextual: an account of an imaginary work of art which can also be understood as a commentary on Barthes's reading of Balzac's *Sarassine.*[18]

The work which Simon describes and discusses clearly falls into Barthes's category of the 'readable' text. It belongs, like *Sarassine*, to the representational tradition. This does not imply a naïve imitation of nature. The two officers depicted within a wide, high room are drawn in accordance with neo-classical conventions ('des anatomies stéréotypées apprises sur l'antique', LesG, 12), just as the room itself is sketched using the conventions of classical perspective. The spectator is required to know these conventions; understanding depends on 'un code d'écriture admis d'avance par chacun des deux parties, le dessinateur et le spectateur' (LesG, 13).[19] Given this knowledge of codes, the spectator can appreciate that, despite appearances to the contrary, this is a finished work. If the room is no more than sketched, if the bodies of the two men are merely outlined in charcoal, if only the faces, and one pair of hands are painted, this corresponds to a sense of hierarchy which places human actions, motives and personality before the body and relegates the inanimate – the room, its furnishing and ornaments, the material world outside – to a distant third place. Practised readers of Simon will here recall his attack on such hierarchies in *La Bataille de Pharsale* and elsewhere, and discern in this configuration of elements an underlying critical tone. In Barthes's terms, this is a 'consumable' work: the shared knowledge of conventions implies a scheme of values common to artist and spectator which the work leaves undisturbed.

Or at least in appearance this is so. For, like Barthes in *S/Z*, Simon also shows that the most conventional of works leaves loose ends that cannot be tied up in a single knot of sense. Viewed closely, the techniques which establish hierarchy also create effects which are 'bizarre', 'unusual'. Styles clash: the classical nudes are surrounded by eighteenth-century furniture. Conventions are inconsistently used: the source of light bears no relationship to the shadows cast by the figures. Above all the work bears traces of the process of its own creation. The older man's phantom hand, almost painted out, shows that the painter hesitated in choosing gesture and moment. This visible trace of arbitrary choice runs counter to the idea that the work could ever be finished or closed.

By implication then Simon is declaring a preference for the kind of text which Barthes decribes as 'writeable', opposed to the 'readable', and superior to it, in Barthes's view, because it challenges received codes and thus stimulates the reader to become, to his or her delight, a producer of meanings. These will be as infinite as the readers themselves and their own constantly changing knowledge of codes and texts. Yet the overlap between Simon's values and those of Barthes is far from complete.

At the simplest level *S/Z* differs from *Les Géorgiques* in that Barthes's work remains a commentary, however open-ended, whereas Simon, while analysing this supposed sketch, also describes, reanimates and transforms it. To put it another way, Barthes's aim is to analyse how meaning is constructed, while Simon is concerned not just with how the picture means, but what it means. This search and discovery of meaning takes two forms, both alien to Barthes. The figure most notably absent from Barthes's book is Balzac, not unexpectedly so. Two years before *S/Z*, in 'La mort de l'auteur' (1968), Barthes had laid definitively to rest the author as controlling principle or source of knowledge. Barthes is concerned not with intentions but with codes of reading. By contrast, the word 'artiste' and its various equivalents – 'dessinateur', 'peintre', 'auteur' – recur frequently in Simon's prologue. Simon is unfailingly interested in what the artist believed, wished or intended:

> Il semble que l'artiste, suivant une sélection personnelle des valeurs, ait cherché, dans la scène proposée, à nettement différencier les divers éléments selon leur importance croissante dans son esprit. (LesG, 14)

> Ceci [. . .] semble confirmer qu'il ne s'agit pas là d'une toile inachevée, mais d'une oeuvre considérée par son auteur comme parfaitement accomplie. (LesG, 16)[20]

However speculative the attempt, signalled in 'semble' and the subjunctive, Simon unequivocally aims to get at the truth of the author's values and intentions.

Secondly and similarly, the characters represented in the sketch – or produced in the writing of it – inspire Simon to speculate about what the men are doing and what they mean to one another. Do the younger man's impassive features mask worry or scorn? Why has he not sat down in the chair seemingly reserved for visitors? What is the significance of the letter (or is it a letter?) which the older man so attentively scrutinises? No conclusions are reached. Indeed the last paragraph becomes a *reductio ad absurdum* of the search for explanation. What was the meaning of the older man's gesture of dismissal? Which leads back to the prior question: was it given before or after he had read the letter? But this gesture is precisely the feature of the sketch which the author had eventually taken pains to rub out. Castles of speculation therefore are being built in the thin air of a next-to-non-existent referent. And yet Simon's curiosity is unbounded and catching. Vivid detail, argued hypotheses prod the reader to ask the same and more questions about motive, intention, relationships, meaning. Simon acknowledges the flimsy basis of evidence, the coded nature of knowledge, and yet invites the reader to search with him for the centre of the maze, where, possibly but improbably, stands Truth.[21]

The prologue denounces conventionalism (yet shows how understanding depends on it), casts representation and reference in doubt, but hankers after the truth and stimulates the reader to search for it. The same is true of the novel's first chapter. It begins by transgressing a variety of conventional codes:

> Il a cinquante ans. Il est général en chef de l'artillerie de l'armée d'Italie. Il réside à Milan. Il porte une tunique au col et au plastron brodés de dorures. Il a soixante ans. Il surveille les travaux d'achèvements de la terrasse de son château. Il est frileusement enveloppé d'une vieille houppelande militaire. Il voit des points noirs. Le soir il sera mort. Il a trente ans. Il est capitaine. Il va à l'opéra. (LesG, 21)[22]

The initial third-person pronoun is like the architect's drawing of the prologue: it has no antecedent, it refers to no 'monument déjà existant'. Nor is it, though anonymous, a fixed pole since, shortly after the passage quoted above, it begins to link events which defy human chronology: a man who was ambassador to Naples under the Directorate runs little risk of being strafed by aircraft fire. To what genre do these brief biographical notations belong? They could be from a curriculum vitae, were it not for the mingling of apparently insignificant and significant details: 'Il est général en chef de l'artillerie de l'armée d'Italie . . . Il voit des points noirs.' And why trifle with the chronological convention by juggling the man's ages – if it is indeed one man – and by using the present tense throughout? Typographically, too, the first chapter quickly breaks with convention. Although according to the most common code, italics signify importance, this text switches between roman and italic type without the slightest change in tone. In short, *Les Géorgiques* differs in one sense at least from what is implied about the drawing described in the introduction: Simon uses familiar codes in unfamiliar ways: unfamiliar, yet not unintelligible, for as one reads *Les Géorgiques* one learns how to read it.

For example, the typography of the first chapter gradually forms a pattern. In the first of the three sections of Chapter 1, roman type is reserved for events in the Revolutionary and Napoleonic periods. In the second section this procedure is reversed: roman type is used for twentieth-century events. Thus typography helps the reader differentiate between various antecedents (they might more appropriately be called 'postcedents') of the initial pronoun. This process is taken a step further when use of the letters L.S.M. and O. imply a further subdivision into three 'postcedents': a Revolutionary general; a volunteer in the Spanish Civil War; and a cavalryman fleeing in the débâcle of 1940. Why then do italics disappear entirely in the third section of the chapter? The effect is to overlay differentiation with a sense of sameness. Thematically, as in Simon's other novels, the reader is invited to view all human lives as variations of the same patterns universally repeated.

Yet there is also another aspect to that sameness. At the beginning of the chapter 'facts' almost overwhelm the reader: biographical data spill over him or her in a disordered flood. But even as the flood is channelled into three streams, one potentially troubling resemblance between the characters becomes apparent: all three are writers:

> Dans le récit qu'il fait des événements O. raconte qu'aux premiers coups de feu un inconnu le prend par le bras et l'entraîne en courant. (LesG, 51)

> il (le cavalier) rapporte dans un roman les circonstances et la façon dont les choses se sont déroulées entretemps [. . .] on peut considérer ce récit comme une relation des faits aussi fidèle que possible. (LesG, 52)[23]

As for the general, the verbs of action which distinguish him at the beginning of Chapter 1 gradually give way to a single repeated verb: 'il écrit', followed by a quotation or paraphrase of his words. The solidity of the biographical 'facts' was deceptive: all that is known about these characters is known through their writings. Simon is not representing these lives, but rather retelling lives already transformed into texts. By abandoning italics he signals that all three characters have the same status: they coexist as fictions in the domain of words.

The first chapter, then, as so often happens in Simon's work, pulls in two directions at once. It betrays a nostalgia for representation, for the precise notation of mood, of perception, of historical or biographical fact. At the same time it undermines the idea of representation. By blurring conventional codes, it draws attention to them, and so carries forward from the introduction the idea, made explicit towards the end of the chapter, that all writing rewrites the previously written.

These same tensions persist in the following chapters. In one sense clarification proceeds apace. Each chapter centres on one or other of the three epochs and three characters: Chapter 2 on the cavalryman in the winter of 1939–40; Chapters 3 and 5 on the general and the long shadow he casts on his family and descendants; Chapter 4 on O.'s experience at the Front and in Barcelona in 1936 and 1937. Techniques inherited from the realist tradition and from Simon's earlier fiction foster the illusion of knowledge. Each chapter has a narrative thread; indeed narrative wrests from description the dominance it had lost in Simon's work since *La Bataille de Pharsale*. The first part of Chapter 2, for example, is not just a nameable but a named narrative sequence, an explanatory account of the 'désagrègation ou si l'on préfère désintegration complète d'une troupe encadrée en quelques heures d'une marche de nuit' (LesG, 94).[24] Chapter 5 blends the reconstructive modes of *La Route des Flandres* and *Histoire*. A narrator – 'le visiteur', 'le garçon' – reconstructs, in memory and imagination, the world of his childhood and the life of his ancestor, based on surviving relics: a ruined farmhouse, the decayed inscription on a gravestone, a heterogeneous cache of letters and papers. The narrative sidesteps and doubles back on itself; yet for every one step back, it takes two forward. At the core of the general's life lies an enigma or rather a series of enigmas which affect later generations. Once Secretary to the Convention and member of the Committee of Public Safety, responsible for promoting and commanding generals, L.S.M. was later dispatched to take charge of artillery in Alsace. Why did a man so powerful lose his power and status? Why was the family's pride in their ancestor tinged with shame? Why had the boy's grandmother preserved his papers, but kept them hidden? In *La Route des Flandres* the enigmas of de Reixach's death, his relationships with Corinne and Iglésia remain intact. The mysteries of *Les Géorgiques*, like those in Balzac's *Sarrasine*, receive plausible, uncontradicted solutions. L.S.M. offends his royalist descendants because he voted for the death of the king. His legitimist brother's death compromised him on every hand: an affront to the family because he had voted for the decree condemning all arms-bearing emigrés to death, a scandal to his fellow revolutionaries because he interceded on his brother's behalf.

The reader gradually gains familiarity then with three differentiated and even hierarchised characters. At the centre of the triptych stands the heroic, tragic, tragi-comical figure of the general resurrected in the particularity of his life and times: the revolutionary commitment of his youth, the unflagging energy of his middle years (as determined to reshape Europe as to order his estate), the feebleness of his old age, his burlesque posthumous adventures: heart removed and bottled in error, grave destroyed by a bulldozer constructing a new road. Two twentieth-century pygmies flank this colossus. The cavalryman shares his experience of war, the Spanish volunteer both war and revolution. But whereas the general commands, and strives, ultimately in vain, to dominate events, both cavalryman and volunteer are of lower status and lesser ambition, caught up in events which undo them.

It has been argued that all that differentiates these characters is nullified by the sameness of their experiences which 'tends towards the flattening of differences, towards the melting down of separate entities and identities'.[25] One might equally argue that the analogies between them contribute to the reader's sense of accumulating knowledge. The stories confirm one another. Each is typical of its times; together, they confirm the universality of the experience shared by their protagonists. L.S.M. and O. participate in revolutions which fail, and ultimately consume their perpetrators. In both war and revolution, as all three characters discover, the strings are pulled by those far from the front line. Nature and war are multiply analogous to one another: each is cyclical and destructive (regimental discipline is undermined by winter, just as it will be six months later by enemy fire); each demands the same qualities of patience and endurance:

> *cet éternel recommencement, cette inlassable patience ou sans doute passion qui rend capable de revenir périodiquement aux mêmes endroits pour accomplir les mêmes travaux: les mêmes prés, les mêmes champs, les mêmes vignes, les mêmes haies à regarnir, les mêmes clôtures à vérifier, les mêmes villes à assiéger, les mêmes rivières à traverser ou à défendre, les mêmes tranchées périodiquement ouvertes sous les mêmes remparts.* (LesG, 447)[26]

What emerges then is, if not a systematic, at least a coherent set of themes, even convictions.[27] Certainly it seems to be with the authority of these convictions that Simon conducts his reading and writing of Orwell's *Homage to Catalonia.*[28] The third section of Chapter 4 deals with O.'s experience at the Front. Orwell's record of this experience corresponds to the perception of war familiar to readers of Simon's previous novels. For those who undergo it, war effects a transition from book knowledge to real knowledge. Reduced to being a creature of basic needs, the individual is stripped of idealist illusions and comes closer to nature. Privation and fatigue, the disorganisation of settled habits, heighten the perceptive faculties. Thus in a sense the experience of war, for all its horror, becomes one of exhilaration and even, as Orwell puts it, of enchantment (LesG, 348). Simon manifestly approves of Orwell's account. However, in most of Chapter 4, his tone has a strongly critical edge. The narrative voice is doubly knowledgeable: familiar with events in Barcelona which Orwell either does not know or fails to acknowledge; filled with the certainty of its own beliefs. Orwell's offence, in Simon's eyes, is to insist on seeing order and logic in the events he had lived through. He compounds this offence by writing his account of Spain in chronological order and coherent sentences, partly to reassure himself, partly for others whom he will consequently mislead as to the truth of what he experienced. All this is the more shocking since Orwell had enjoyed all the opportunities Spain provided to have his eyes opened. Yet, when it came to writing, he refused to take the path traced by protagonists in previous Simon novels. Unlike Louis in *Le Tricheur* or Bert in *Gulliver*, O. fails to acknowledge the essential disorder of life; unlike the student in *Le Palace*, he refuses to recognise that the revolution was stillborn; unlike Louise in *L'Herbe* he does not scorn syntax which imposes false coherence on experience which is chaotic; unlike the narrator in *La Corde raide*, he does not recognise and resist the temptation to write with an eye cocked towards his readers. In short, the problem with O., in Simon's view, is that he is too much like Orwell, not enough like Simon.

Celia Britton has also discussed the authority of Simon's voice in this chapter of *Les Géorgiques*. She implies that it is unusual in Simon's work to find such 'a discourse of knowledge, fulfilling the traditional didactic function of realism'; it contrasts with 'the hesitant and exploratory nature of most of Simon's writing'.[29] While not dissenting from this view in general, I would argue that discourses of knowledge appear quite frequently in the novels, even within the limited field of characterisation. Many such discourses are handled ironically. There are, for example, the purveyors of false knowledge, whether confident in their beliefs, like the priest in *Le Tricheur* and the solicitor in *Le Vent*, or clinging to them hesitantly, like Pierre in *L'Herbe* and *La Route des Flandres*. Simon's doomed early heroes, from Louis in *Le Tricheur* to Bernard in *Le Sacre du printemps*, form a second category. Their claims to knowledge are shown to be equally false, whether they hope in revolution or choose to believe that life is 'une suite d'épisodes réflexes découlant les uns des autres' (SP, 24).[30] Their fate is to be stripped of these beliefs by the adventures they undergo. O. is a belated example of this type, frustrating to Simon because he refuses to conform to the familiar scheme. But there is also a discourse of knowledge involved in the ironic presentation of such characters. To present their beliefs as illusions is to proclaim yourself the possessor of a superior truth, even if what you claim to be affirming is the absence of all certain knowledge.

The paradoxically strident critique of Orwell, almost a throwback to the novels of the early 1950s, is the strongest affirmation of truth in *Les Géorgiques*. Elsewhere the effect of intertextual references is to soften the edges of clarification. The intertexts open *Les Géorgiques* out to multiple meanings. What, for example, are the resonances of the title? Georgics, both traditionally and in Virgil's version, are a guide for farmers, a treatise on the work of the land, season by season. Here lie the most obvious parallels between Simon and Virgil. From all over Europe L.S.M. sends letters to Batti, the servant who manages his farm. They are as full as Virgil's *Georgics* of instructions and recommendations about raising crops, breeding horses or tending vines. Stylistically they share a similar predilection for the imperative mood. One may guess that these characteristics of his ancestor's letters first put Virgil into Simon's mind as a point of reference and comparison. Other resemblances followed.

Thematically both Virgil and Simon play on an analogy between the work of the land and the work of war. In *The Georgics* the analogy is no more than hinted at and the war work of Augustus reaches the writer in his peaceful countryside as a distant echo. Simon reverses Virgil's perspective: battles, wars and revolutions are in the foreground of *Les Géorgiques*; from the frontiers of empire L.S.M. longs for the life of the land and the peaceful turning of the seasons. And as we have seen, Simon makes explicit the analogy: the land and war require the same qualities.

Although L.S.M. is said to be a lover of Virgil, there are no quotations from *The Georgics* in the novel, relatively few explicit, incontrovertible references to characters or motifs. This shadowy presence is precisely what stimulates readers to seek connections between Virgil's and Simon's *Georgics*. For Mary Orr, Simon's version is an admiring but 'ironic adaptation' of Virgil.[31] Virgil describes a bucolic utopia; Simon's treatment of the seasons, cultivation, men, animals, and even Italy mocks Virgil's optimism. For Michael Evans, 'Simon's novel weaves its own allegories out of the Orpheus myth', allegories of transgression, imitation and reflexivity.[32] The effect is partly ironic and deflationary, for example in Simon's disenchanted account of a performance of Gluck's *Orpheus and Eurydice*; but in part a matter of parallels: the Orphic gaze 'doomed to dissipate the very past it seeks to revive'.[33]

Another possible interpretation is to see *The Georgics* as an image standing for antiquity. Virgil's poem generates a host of references to Latin and Greek civilisation which expand Simon's themes of tradition and innovation, permanence and change. These references establish the ironic paradox of an age which was determined to innovate (for what could be more innovatory than carrying out a revolution and executing a king, not to mention stripping the months of their classical names and turning the calendar back to zero?), an age determined to innovate, yet which in many respects modelled itself on antiquity. And not just in matters of fashion (the general's wife – 'Julie', 'Briséis' – dressed her hair 'à la grecque'), but also in the way it conceived of itself, both in private life, as witness the references to Greek and Sparta in the inscription L.S.M. chose for his first wife's tomb, and also in the vocabulary and forms of government: tribunal, consulate, empire. Hence also, modelled on the initial drawing, and apt illustration of the circular influence of life on art and art on life, the general's marble bust which represented him 'non pas revêtu de son uniforme de général aux pesantes dorures mais les épaules drapées d'une toge à l'antique [. . .] cette effigie de tribun' (LesG, 196–7).[34] This whole series of images contrasts with others applied to the gypsies who surround the young boy in the foetal darkness of the cinema: 'c'était comme une sorte de grouillement confus, le confus réveil, le confus remue-ménage d'une horde barbare' (LesG, 212).[35] These barbarian gypsies spring fully-formed from the depths of history, and incidentally from the pages of *Le Vent*. They represent all that is changeless: 'la délégation vivante de l'humanité originelle, inchangée, les spécimens inaltérés et inaltérables, rebelles aux siècles, au progrès' (LesG, 208).[36] They undermine by contrast L.S.M.'s idealism and commitment to change which also stand, ironically, on a classical foundation, 'la tranquille détermination puisée dans la lecture de ces auteurs latins' (LesG, 197).[37]

The forms of the past survive in life and also in art. These same themes emerge in different guise from Simon's treatment of one of the few narrative elements he borrows directly from Virgil: the legend of Orpheus and Eurydice. This forms one of the innumerable themes which link times, characters and places: the general meets his first wife at a performance of Gluck's *Orpheus and Eurydice*; his descendant is taken as a boy to the same opera, and later, as a soldier at the front, he hears snatches of it on the radio. Yet Simon's choice of Gluck's version of the story of Orpheus is more than a convenience. It emphasises that, although the legend has a kind of permanence, it exists in forms which, always related, are constantly changing. Simon quotes Gluck, who had adapted Virgil, who in turn was retelling a Greek story. The legend passes from genre to genre – narrative, poem, opera, novel – and from language to language – Greek, Latin, Italian, French. The artist then is not a creator, but an adaptor. All his texts are second-hand because each work is inescapably a transformation of previous works.

More than any other, Chapter 3 of *Les Géorgiques* gives a sense of that transformation in action, the play of text on text which sets representation at an infinite distance. As boy and man, L.S.M.'s descendant strives in imagination to reconstruct his ancestor's life. The key relationship here is that of fascinated to fascinator, as in *Le Vent* or *L'Herbe*. In structure, the chapter resembles *Histoire*. Surviving fragments of the ancestor's life – 'l'amoncellement de paperasses, de vieilles lettres et de registres' (LesG, 193)[38] – emerge, like the postcards in *Histoire*, from a long untroubled resting-place in the family house. The resemblance becomes more pointed when reconstruction of the descendant's childhood brings apparently familiar characters out of the family cupboard: the grandmother who is the head of the household, and the young boy's mentor and informant Uncle Charles. Even so, the parade of familiar forms is not limited to *Histoire*. The descendant's visit to the peasant occupiers of L.S.M.'s decayed 'château' has the same burlesque overtones as Montès's visit to his father's farm in *Le Vent*. The old woman in *Les Géorgiques* combines elements from *L'Herbe*. Like Marie she endures her final agony under the guard of a nurse of vaguely fabulous appearance and attributes (LesG, 192–3); like Sabine she confers 'une dimension apocalyptique à tout ce qui pouvait constituer quelque motif ou prétexte de souci' (LesG, 195);[39] and her male companion, though named Charles, is also described as 'un impassible homme de pierre' (LesG, 203).[40] The general himself appears sometimes as a figure of flesh and blood, sometimes as an image in marble. His metamorphoses recall similar transformations in *La Bataille de Pharsale*. Thus the forms which appear in Chapter 3 – fiction, characters, motifs – are largely forms inherited from Simon's work and blended in new shapes and patterns.

In addition, however, the chapter invites the reader to see links with other texts. One of these has already been considered; it is largely in this chapter that references to Virgil's *Georgics* and to antiquity abound. In listening for yet other voices one can at the same time observe how the aesthetic of representation is challenged here. For to aim to reconstruct the past is one of the possible particular forms of that aesthetic. Does Simon suggest that L.S.M.'s descendant succeeds in recapturing the reality of the general's life? 'Et il semblait au garçon qu'il pouvait le voir':[41] this phrase from near the end of the chapter (LesG, 250) might lead one to believe success had indeed been achieved, were that belief not shaken by a variety of intertextual references. First the phrase itself is very familiar to practised readers of Simon: used by the narrator in *Le Vent* and Georges in *La Route des Flandres*, it bears ironic overtones of hope and confidence misplaced in the power of 'imagination and an approximate logic' (V, 9–10). Further, the comparison which follows the introductory phrase is itself an admission of defeat: 'il semblait au garçon qu'il pouvait le voir: le corps puissant, musculeux, commençant à s'alourdir, nu, comme ces dessins copiés sur l'antique'.[42] Just like the portraits of the Revolutionary age, the boy's view of his ancestor is presented as a conglomerate of borrowed familiar forms. The L.S.M. of *Les Géorgiques* is not a unique individual directly represented, but a rearrangement of clichés. Finally, all hope of touching reality directly is undermined by another impression of 'déjà lu'. The third chapter of the novel is not simply a reworking of Simon or of Virgil, but also of Faulkner.

'Descendre de Faulkner? Ça ne me gêne pas. Nous descendons tous de quelqu'un.'[43] Simon has frequently expressed his admiration for Faulkner. It was particularly evident in his novels of the late 1950s from *Le Vent* to *La Route des Flandres*, so much so that critics, old style, spoke in terms of influence, and *Tel Quel* listed among the clichés of 1960: 'Claude Simon: le Faulkner français'.[44] In Chapter 3 of *Les Géorgiques* Simon pays conscious, affectionate homage to *Absalom, Absalom!* The references are most explicit at the beginning and end of the chapter, as if to put the remainder of it within parenthesis. It begins, as does *Absalom, Absalom!*, with the description of an indomitable old woman, dressed in black, faithful to a vanished past: this survivor from *Histoire* is also a version of Miss Rosie Coldfield. When 'le visiteur' makes his way to a decaying and abandoned tomb, 'sous le mélancolique et silencieux bruissement d'une pluie d'automne' (LesG, 197),[45] he is retracing the squelching footsteps of Quentin and Mr Compson. All three are on the track of a man who dwarfs and fascinates them. L.S.M., like Sutpen, is a man of superhuman energies whose rise and fall mirror the ambitions and passing of a heroic age. Stylistically the resemblances are evident throughout the chapter in the convoluted prose; syntactical order is barely maintained by a variety of parenthetical devices. Simon's homage culminates in the gently humorous pastiche of Faulkner's manner at the end of the chapter. A garrulous, disillusioned old man and a naïve, astounded youth piece together the fragments of a family secret which may or may not explain the conduct of a man long dead:

> 'Parce qu'il avait un frère . . .', et le garçon: 'Un fr . . . Quel frère?', et l'oncle Charles: 'Légalement et biologiquement. Oui. Parce qu'il est convenu de donner ce nom aux produits de deux embryons issus des mêmes glandes mâles et grandis dans le même ventre. Sauf qu'ils se ressemblaient à peu près comme un négatif photographique ressemble à l'épreuve tirée. C'est-à-dire exactement pareils et exactement contraires . . . ', et le garçon: 'Alors il avait un frere? Mais pourquoi . . .',

> et l'oncle Charles: 'Tu veux dire: pourquoi est-ce qu'on n'en a jamais parlé? Eh bien voilà: précisément!' (LesG, 255–6)[46]

Faulkner speaks through Simon in the repeated 'parce que', in the ironic definition of brother and in the manipulation of the voices. The flavour of melodrama is pure Faulkner, straight from a typical chapter ending in *Absalom Absalom!*

> He (Quentin) couldn't pass that. He was not even listening to her; he said, 'Ma'am? What's that? What did you say?'
> 'There's something in that house'.
> 'In that house? It's Clytie. Dont she –'
> 'No. Something living in it. Hidden in it. It has been out there for four years, living hidden in that house.'[47]

In both cases a bombshell of new information provokes an incredulous response and a staccato exchange of words. The new information raises new questions. Why has no one ever spoken of the brother? Who has been living in the house for four years? Both questions will be answered; but Simon plays his last Faulkner card by keeping the reader in suspense for the next 200 pages.

In one sense *Les Géorgiques* can only be read intertextually. As readers, we have no other choice: we are what we have read. Literature is like the proverbial Spanish inn: you find what you bring to it. The real question is one of degree. To what extent is a knowledge of specific works required before you can understand or take pleasure in *Les Géorgiques*? How much does it matter if a reader is unable to connect O. with Orwell, or indeed with Simon's previous use of the same letter to stand for a point of view in *La Bataille de Pharsale*? If you haven't read Virgil's *Georgics*, are you excluded from Simon's? To argue for the indispensability of such previous reading appears to stem from the conviction, nourished by Barthes, that no text is closed: sense arises from the play of one text against another. But, more deeply, such a view comes close to implying the opposite; that in the end meaning is available only to those who, going the long way round, amass sufficient knowledge to grasp it. What comes first in this novel, however, is not O.'s relationship to Orwell, who is never mentioned by name, or Virgil's to L.S.M. or to Simon. What matters most are the relationships, similarities, comparisons, contrasts, between three figures, L.S.M., O., the cavalryman, three experiences of war and revolution, and three ways of writing about that experience. More than that, in *Les Géorgiques* Simon makes sufficient use of the traditional codes of realist fiction in twentieth-century mode – character, story, narrative sequence, resolved enigmas – to make his novel accessible and pleasurable to general readers whose cultural baggage does not include specific named texts or the whole prehistory of Simon's fiction. If you can take pleasure in *The Sound and the Fury*, you can enjoy *Les Géorgiques*.[48]

On the other hand, *Les Géorgiques* does insistently indicate that the world of representations outside itself, visual as well as written, is somehow responsible for its own origins. Clearly knowledge of some of these must enrich the reader's pleasure in the novel. My own reading, limited to written intertexts, has dwelt on three kinds of relationship. Both Virgil and Orwell are sufficiently signposted and treated to make it clear that Simon had read and was using them in *Les Géorgiques*.[49] Quotations from Michelet and the discussion of Strachey fall into the same category.[50] Faulkner, it has been argued, is a more doubtful case.[51] A mere subjective association, a bee in my bonnet? The case has to stand or fall on the strength of the critic's argument and evidence. I am comforted by knowing that Simon had read *Absalom, Absalom!* and previously used it in his novels. Chapter 2 of *Les Géorgiques* bears a somewhat similar relationship to Balzac: there is more than a hint of parody in Simon's declared intention to explain a descent into chaos. Yet a third type of intertextual relationship is exemplified by the use I have made of Barthes. I introduced the comparison with *S/Z* to elucidate *Les Géorgiques*. I make no claim that Simon uses it as an intertext. The pertinence of the comparison is that both Simon and Barthes are reflecting in not dissimilar ways on how works of art are produced and interpreted.[52]

This use of Barthes illustrates the most pervasive kind of intertextuality in contemporary critical writing. What academic critics bring to novel or poem or play is of course a knowledge of other novels, poems and plays; but, as much and often even more, their knowledge of other critics. No doubt this is originally an inheritance from the nineteenth century. Literary studies are shot through with scientism and the belief in progress. Convention requires us to set out our arguments, and footnotes, as if we were building on the work of our predecessors and contributing our grain of sand to the heap of human knowledge. Despite its very different philosophical presuppositions, the explosion of critical theory over the last thirty years has exacerbated the trend towards critical intertextuality. Hence, in large part, and in no small part reluctantly, what this book is about:

not interpreting Simon's novels, but interpreting interpretations of Simon's novels. I would love to make the definitive statement, tell the truth about Simon; but as a child of my times, I have imbibed the thin milk of relativism, even if I don't like its taste.

The danger of such academic criticism, as I see it, is that, far from drawing new readers to Simon, we tend to drive them away. Have I, however unwillingly, set up a new barrier: no access to *Les Géorgiques* except through Barthes? Perhaps the best guide to intertextuality in *Les Géorgiques*, and to Simon's model reader, is given in the description of Batti puzzling over her master's letters:

> les missives cachetées de cire qu'elle ouvrait, déchiffrait ou plutôt décryptait, essayant de voir dans ce qu'il appelait la division bleue la division verte, la division rose, ces peupliers, ces acacias, ces champs, ces vignes expédiés en quelque sorte par la poste sur des rectangles de papier couverts de petits signes à partir desquels (à la maniere de ces microscopiques fleurs japonaises qui, précipitées dans l'eau, se gonflent déploient des corolles insoupçonnées) se matérialisaient à nouveau la terre exigeante, les coteaux, les vallons tour a tour verdoyants, roussâtres, desséchés ou boueux sous les ciels changeants, la lente dérive des nuages, la rosée, les orages, les gelées, dans l'immuable alternance des immuables saisons. (LesG, 462)[53]

Like Batti, the reader is constantly confronted by codes. Although other texts have often resonated in Simon's novels, never before has the transformation of existing forms and texts been so willingly proclaimed and practised as a principle of construction. In this example, the flowers are as Proustian as they are Japanese, and they unfold into multiple references to Combray and the Bois de Boulogne. Elsewhere Simon juggles with elementary codes or attacks the whole temper of a work by rewriting it (*Homage to Catalonia*), or offers homage through pastiche (Faulkner's *Absalom, Absalom!*). A marked effect of these varied references is to warn the reader that 'all is untrue'. L.S.M. is not Simon's ancestor, O. is not Orwell, neither 'le cavalier' nor 'le visiteur' are Simon. In fact, Simon is manipulating signs. His aesthetic is the very reverse of the last-ditch realism proposed in different ways by Truman Capote's *In Cold Blood* or Thomas Keneally's *Schindler's Ark*: his retold lives are strongly, avowedly fictional; the reader is constantly reminded that this is so.

Nevertheless the effect of intertextuality in *Les Géorgiques* is not simply to make representation more problematic. The presence of other texts increases the distance between words and reality; but through their resonances and echoes such texts multiply perspectives on the elusive real. It is as if Simon is using intertextual play to free himself from the straitjacket of a certain theoretical orthodoxy: by reintroducing variety of style, manner and technique, intertextual play helps involve the reader in almost forgotten ways. When Batti has deciphered the code of colour and writing, the material world seems to present itself to her in all its varied, recurrent, unfailing sensuousness. Such also is the experience of reading *Les Géorgiques*. The rearrangement of clichés does not produce cliché; rearranged, refreshed, they invite the reader to participate imaginatively and emotionally in a game of conscious invention, or rather, reinvention: to reinvent from the pages of Simon's earlier novels both 'le cavalier' and 'le visiteur' and their tragicomical obsessions with disintegration and decay; to break Orwell's self-image and recast it in a different mould; to re-create from fragmentary papers the haunting figure of L.S.M. And beyond these characters all the signs point to another whose intertextual adventures are just as moving. Lucien Dällenbach has suggested a final twist to the meaning of the title:[54] *Les Géorgiques* is a new version, or perhaps rather three new versions, of the legend of Georges, named as a character in *L'Herbe* and in *La Route des Flandres* and who appears in different guises as protagonist in all the novels up to *Les Corps conducteurs*. What is true of the prologue to the novel is also true of the novel's total effect: the intertextual here does not displace the inter-subjective, but engages the reader in a search for contact with a subjectivity which flits between, slips away between the lines. *Les Géorgiques*, reinventing a life spent in books and on battle fronts, retells the story of a character who is both fact and fiction: Claude Simon.

5

Autobiographical fictions: *La Corde raide* to *Les Géorgiques*

The idea that you can write the story of your life is provocatively humanist. It implies the existence of a conscious coherent self which uses language to represent lived experience. Historically, writers of autobiography have often aimed even higher. Rousseau claimed that in telling the truth about himself he provided a first piece of comparative evidence for the study of mankind.[1] The intellectual climate of the 1960s and 1970s was profoundly hostile to such ideas and ambitions. Sceptical about human nature and universal truth, the theorists of deconstruction attacked the concept of the unified self and the idea that language could be used to give access to it. Where autobiography, conventionally understood, posits the authority of the writer over the text, Derridians and Lacanians were united in affirming that the full presence of self in text is a mirage: writing is the trace of an origin which is lost. For theorists of intertextuality that origin lay in previous texts or, more generally, in the labyrinth of social and cultural codes which deny access to any ultimate extra-textual reality. For Ricardou, whose writings formed the micro-climate in which Simon and Simon criticism moved in these years, autobiography combined the double illusions of expression and representation, expression of the self and representation of a pre-textual reality. When Ricardou's reading of *La Bataille de Pharsale* was challenged by a reader who claimed that the novel originated in part from Simon's personal experience, Ricardou brilliantly demonstrated how the words *Gabbia d'Oro* interacted with other components of the text. Whether Simon had ever spent a night in a hotel of that name was irrelevant; what mattered were the transformations operated linguistically within the novel.[2]

Yet by the mid-1970s, when deconstructive criticism was still at its height, a counter-current, favourable to autobiography, was already growing in volume. Critically, it began with Philippe Lejeune, whose work has spanned the gulf between the official, still largely humanist culture of the French universities and the lively terrorists on its margins. *L'Autobiographie en France* (1971) was a study of genre, hence intellectually respectable to both cultures, although humanist in its assumption that autobiography transmits experience. *Le Pacte autobiographique* of 1975 was more theoretical and formalist in tone. The author of an autobiography, Lejeune affirmed, makes a pact with the reader to tell the truth. But that pact is sealed (at least on the author's part), in that the name given on the title-page is the same as that of the protagonist. By emphasising this formal criterion, Lejeune seeks to play down the tricky questions of intentionality, sincerity and truth. In the introductory essay to *Moi Aussi*, first published in 1983, Lejeune criticises his earlier formalism. The theoretical assumptions on which the idea of autobiography is based are all highly questionable, and yet as readers, he asserts, we all make them. In theory autobiography is impossible, in practice it exists.[3] This is the core of Lejeune's thought. As a theorist, he is consciously naïve. The force of his example has lain in the breadth of his learning and the subtlety of his analyses of specific texts. His studies of Sartre, Leiris and, more recently, Perec, have gained conviction for his pragmatism.[4] By drawing on biographical facts and psychoanalytic theory, by close textual study of texts in draft and published form, Lejeune has shown that autobiographical criticism, like autobiography, works.

On the hazy frontier between critical theory and creative writing Barthes set the new tone in his *Roland Barthes* by Roland Barthes in 1975. Barthes shadow-boxes with the idea of the autonomous self, controlling its own language, transmitting truths to its readers. There is a similar, typical playfulness to Robbe-Grillet's part-autobiographical volumes, *Le Miroir qui revient* (1985) and *Angélique ou l'enchantement* (1988). They arose from an invitation to write a book parallel to Barthes's, in the same series. Their collective title, *Romanesques*, proclaims their continuity with Robbe-Grillet's novels. The fantastic mingles with the factual so as to confound all possible readings: readers looking for fact will find fiction, and vice versa. Yet this extension to Robbe-Grillet's techniques for keeping sense in suspension is at the same time a reaction against the writer's demotion to mere 'scriptor', the product of language. 'Je n'ai jamais parlé d'autre chose que de moi.' In these opening words of *Le Miroir qui revient*, first published in 1978,[5] Robbe-Grillet fires a triple salvo designed to distance himself provocatively from Ricardou, to affirm a proprietorial authority over his work, and to invite an ambiguously autobiographical reading of his novels.

The reaction came with greater force amongst others who had been shaken but not been broken by the prevailing winds. In *Le Livre brisé* Serge Doubrovsky recalls with mock anguish the Terror of the later 1960s:

> S'exprimer est un terme obscène. Si on l'emploie en public, on risque de se faire écharper. Devant Robbe-Grillet ou Ricardou, on signe son arrêt de mort. L'auteur est mort, le Dieu de la critique ait feu son âme. C'est le langage qui parle tout haut, tout seul, personne ne parle, il ne dit rien, sans source, sans origine, il se déploie en grandes formes symboliques.[6]

By 1977, however, following the latest twist in Barthes's career and Robbe-Grillet's partial apostasy, Doubrovsky was bold enough to coin a new term, 'auto-fiction'. In works such as his own novel, *Fils*, the author will engage himself to tell only verifiable facts. Author, protagonist and narrator will share the same name, and the narrator will use the first person. But, within these limits, the author will be free to employ all the artifices of language and fiction.

Sarraute's *Enfance* (1981) breaks these rules. Two anonymous narrative voices construct the minute psychological dramas which play beneath the surface of everyday human encounters. Sarraute calls these tropisms; they are familiar from her novels. Yet this work has a new, strongly autobiographical flavour. Ann Jefferson has argued that, by grounding these tropisms in experiences of childhood, *Enfance* retrospectively validates Sarraute's consistent affirmation that tropisms have a pretextual origin. They persuade us that what Sarraute said about them in the novels is true. But why should readers believe Sarraute when she attributes such experiences to childhood? Ann Jefferson replies in terms of genre and reader expectation: *Enfance* benefits from '[T]he generic presupposition that autobiography is the retrospective account that someone gives of their own existence'.[7] This analysis glosses over an important aspect of *Enfance*. Would readers find the retrospective validation of the novels convincing had Sarraute set her childhood memories in, for example, a small town in Texas? Some would, undoubtedly. The narrators' double act, their scrupulous concern for truth, adds a new technique of persuasion to the range previously available to writers of autobiography. But for many readers the autobiographical presupposition comes into play only because the narrated episodes from the narrator's childhood cohere with what they know about the life of the author. Natasha shares with her distanced *alter ego*, Nathalie, a troubled cosmopolitan childhood, torn between Russia and France, between father, mother and stepmother. Of course such readers' knowledge of Sarraute's life is also in most cases purely intertextual: what I know about Sarraute, I have read. Even so, *Enfance*'s claim to authenticate the novels is based ultimately on an appeal to what convention supposes is extra-textual reality. In practice, Sarraute and the reader conspire to believe in a continuing self which uses language to convey truth about its past and the nature of reality.

From *La Bataille de Pharsale* in 1969 to *Leçon de choses* in 1976, Simon's successive novels were largely read and to no small extent written in a manner in tune with the Ricardolian times. Nothing could have been further from critics' minds than what Anthony Pugh, writing in 1982 about *Histoire*, called 'the autobiographical element' in Simon's work.[8] Reviewers and critics of *Leçon de choses*, for example, treated it almost universally as a Ricardolian work, almost an exercise in the elaboration of a novel from the stimulus of an initial description. In so doing, they were no more than taking their cue from the short initial section, entitled 'Générique'. Of course, the intratextuality of the novel – Ricardou's term for the links between works of the same author – was clear to see. Its themes – war, violent death, procreation, destruction and re-creation – were characteristically Simonian. Many of its episodes were visibly reworked from previous novels. But at the time it would have been very bad taste to remark, for example, that the massacre at the railway cutting, recounted with humorous verve by the older mason, not merely transfers to a new persona similar episodes of *La Route des Flandres* and *La Corde raide*, but in some sense owes its origins to an experience which Simon underwent with the First Squadron of the 31st Regiment of Dragoons near the hamlet of Lé Fontaine on 12 May 1940.[9] What indeed could be the significance of such an origin to a text which was solely an adventure in language?

Publication of *Les Géorgiques* in 1981 abruptly changed the climate in which Simon's novels were read. Reviewers immediately commented on the new novel's autobiographical basis. Interviewers asked Simon about it. His replies combined encouragement with dissuasive caution: 'A partir de *L'Herbe* tous mes romans sont pratiquement autobiographiques.'[10] On the other hand he warned that the character described in *Les Géorgiques* as ' "l'aîné des garçons", ce n'est pas moi, je veux dire un portrait fidèle de moi, exhaustif aussi; c'est impossible'.[11] The award of the Nobel Prize in 1985 encouraged the new mood. Not merely did it make Simon a public figure, part of France's national heritage – to the surprise, it must be said, of most of his fellow countrymen – it also generated more interviews and articles about his life, which in turn informed the reading of his novels. Indeed, the most striking passage in Simon's reception speech to the Swedish Academy is the few lines in which, summing up his early life, he simultaneously describes the matter of his novels:

> Je suis maintenant un vieil homme, et, comme beaucoup d'habitants de notre vieille Europe, la première partie de ma vie a été assez mouvementée: j'ai été témoin d'une révolution, j'ai fait la guerre dans des conditions particulièrement meurtrières (j'appartenais à l'un de ces régiments que les états-majors sacrifient froidement à l'avance et dont, en huit jours, il n'est pratiquement rien resté), j'ai été fait prisonnier, j'ai connu la faim, le travail physique jusqu'à l'épuisement, je me suis évadé, j'ai été gravement malade, plusieurs fois au bord de la mort, violente ou naturelle, j'ai côtoyé les gens les plus divers, aussi bien des prêtres que des incendiaires d'églises, de paisibles bourgeois que des anarchistes, des philosophes que des illettrés, j'ai partagé mon pain avec des truands, enfin j'ai voyagé un peu partout dans le monde . . . et cependant, je n'ai jamais encore, à soixante-douze ans, découvert aucun sens à tout cela, si ce n'est, comme l'a dit, je crois, Barthes après Shakespeare, que 'si le monde signifie quelque chose, c'est qu'il ne signifie rien' – sauf qu'il est. (DS, 24)[12]

Paradoxically, the passage is both a denial of faith, whether religious or humanist, and an affirmation of the unity and continuity of the self: the same 'I' holds the same views through the years. The effect of such writing round the novels has been to set them in what Lejeune calls an autobiographical space and to encourage re-readings of the earlier novels in this new perspective. What I am going to argue here is that Simon's work can be seen as the hesitant realisation of an autobiographical project.

The origins of this project are to be found in one of Simon's early works, his only text of any length not to be subtitled 'novel'. *La Corde raide*, published in 1947, is a book of memories and reflections which gives every indication of belonging to the genre of autobiography. 'Je suis un homme qui essaie de vivre, je suis tout à cette difficulté de vivre, je cherche ce qui peut m'aider à continuer et pour ça il faut que je trouve du solide sur quoi on peut compter'. (CR, 73).[13] The questions raised in this passage run through the work. Who am I? How did I become who I am? How can I go on from here? To help explain who he is now, and to prepare himself for the future, the narrator/author seeks solid foundations in the past. But though the questions are clear, the answers are much less so. Philosophically, the writer has made certain choices: the self which he constructs here has given up the certainties of religious faith and rationalist humanism. Moral systems simply cover up the essential, 'les trois ou quatre besoins fondamentaux, comme coucher avec des femmes, manger, parler, procréer, pour lesquels les hommes sont faits et dont ils ne peuvent se passer' (CR, 54).[14] Even the absurd is rejected as a 'substitute value': 'Dire que ce monde est absurde équivaut à avouer que l'on persiste à croire en une raison' (CR, 64).[15] Ten years later Robbe-Grillet was to take up this same theme: 'le monde n'est ni significant ni absurde: il *est* tout simplement'.[16] Yet the parallel with Robbe-Grillet is less striking than the continuity in Simon's own view. When in 1985 in Stockholm, Simon affirmed that 'si le monde signifie quelque chose, c'est qu'il ne signifie rien – sauf qu'il est', he was reiterating what he had advanced in *La Corde raide* thirty years previously.

How then can one face up to a world empty of meaning? Unlike many literary autobiographies, *La Corde raide* is not the story of how a writer discovered his vocation. In 1947 Simon had little claim to be a writer. His only published work was *Le Tricheur*. He had tried rather to become a painter. In so far as he sketches a possible vocation, it is in visual terms: 'Je n'explique pas, je constate, et je me borne à raconter ce que j'ai vu' (CR, 138).[17] Simon further explains what this means by reference to painters. He upbraids those who have gone beyond this brief, those who, seeing the world through particular spectacles, have expressed 'une vision idéale du monde, volontairement limitée et arbitraire, chacune de ces visions du monde étant basée sur une conception, ou plutôt une morale de l'univers visible, ou même, ce qui est plus grave pour des peintres, une morale tout court' (CR, 66–67).[18] Cézanne alone, on the other hand, succeeded in capturing 'un univers pour la première fois démuni de poteaux indicateurs. Si totalement dépouillé de tout, excepté de vérité et de cohésion, que pour la première fois s'offrait, dans sa totale magnificence, sans commentaire ni restriction, le monde visible et, à travers lui, le monde tout court' (CR, 117).[19]

By restricting himself to tell what he has seen (perceived, felt, experienced?), Simon hopes to describe the world as it is. These aims, imprecise as they are, can be followed in *La Corde raide* but they lead to an impasse. For one thing, the determination to avoid 'truquages' (falsifications, artifices) takes Simon a long way from his point of departure. To be true to memory, *La Corde raide* is chronologically confused: fragments of stories, memories, discursive passages mingle, interrupt, recall one another. More than this is at stake, however, as can be seen by comparing the beginning and the end of the work:

> Autrefois je restais tard au lit et j'étais bien. Je fumais des cigarettes, jouissant de mon corps étendu. [. . .]
>
> A Paris dans l'encadrement de la fenêtre, il y avait le flanc d'une maison, un dôme [. . .] en le regardant je pouvais voyager et me souvenir des matins où l'on se réveille dans des chambres d'hôtel de villes étrangères. (CR, 9)

This very Proustian beginning introduces a self which dominates its memories, summons, arranges, and enjoys them. The sense of well-being attributed to the character corresponds to the serenity of the enunciation: the imperfect tense holds the past at a safe distance. Only the opening word, 'autrefois', strikes a disquieting note: serenity and well-being belong to the past. That note grows in volume throughout the text which is progressively invaded by the present – the present of memory ('Je me rappelle'), of reflections about life, of conversations with an imagined reader. Towards the end, these invasions definitively disrupt the harmonious unity between character and narrator. Memories, perceptions, thoughts assail the self, split it apart:

> Les bruits et les couleurs se mélangent. [. . .] Tout s'embrouille et s'interpénètre. [. . .] A cause de tout ça, je ne suis pas moi. (CR, 169–70)
>
> Autant chercher à retenir l'eau dans ses doigts. Essayez. Essayez de vous chercher. 'Je est un autre.' Pas vrai: 'Je est d'autres.' D'autres choses, d'autres odeurs, d'autres sons, d'autres personnes, d'autres lieux, d'autres temps. (CR, 174)[21]

At the end of *Le temps retrouvé*, Proust's narrator has become the master of his past; he dominates it as if perched on stilts. The narrator of *La Corde raide* sways precariously on his high wire, at the mercy of a continual present:

> [. . .] Mon sujet n'attend pas.
> Quel est donc votre sujet?
> Une course de vitesse
> Comment cela?
> Des gens et un tas de choses, des odeurs, des heures, des idées, des figures qui courent, et moi au milieu d'eux, à en perdre haleine, pour me maintenir à leur hauteur.
> Vous voulez dire que si vous vous arrêtez, vous ne saurez plus de quoi vous vouliez parler?
> On ne sait jamais de quoi on parle avant d'en parler. (CR, 177–8)[22]

In this race, the self is constantly dissolving and reforming itself, in language.

Both in form and in the logic which leads from memory to the present of writing, *La Corde raide* anticipates Simon's novels of the late 1950s and 1960s. Yet the logic here is not rigorously carried through. Cézanne is praised for having represented 'the visible world and, through and beyond that, the world simply as it is'. Simon sets the same aim for himself. But this form of representation, like any other, itself implies philosophical preconceptions and aesthetic choices, in short, 'an ethic of the visible world'. The parallel with Robbe-Grillet is instructive. When he proclaimed in the mid-1950s a similar wish to escape 'the old myths of depth' by favouring the visual sense, unencumbered by emotive language, critics were soon to remark that the 'cleansing power of sight'[23] was severely limited: accumulating, organised descriptions, no matter how apparently objective, led readers into a world of near psycho-pathological obsession.[24] In *La Corde raide* Simon favours visual description much less radically than Robbe-Grillet in *Le Voyeur* or *La Jalousie*. On the contrary *La Corde raide* is highly discursive. Simon conceptualises his themes in philosophical and metaphysical reflections, comments on the significance of the memories and anecdotes he relates. Placed in these contexts, the fragments of story tend to become illustrations of the author's ideas, chosen to convey his point of view. The old uncle demonstrates the isolation of impending death; the prison camp shows the possibility of anarchic liberty in the midst of apparent constraint. An account of four cavalrymen regaining their own lines ends abruptly to escape the expected code of completion. 'Ce genre d'histoire sans commencement ni fin, le public n'aime pas ça' (CR, 89);[25] but this commentary in itself turns the incident into an illustration of the desire to break that code. Hence the nature of the impasse in which Simon finds himself in *La Corde raide*. He is dissatisfied with received forms; he wants to see life afresh, not coloured or limited by inherited or preconceived interpretations. This drives him towards a new form of autobiography: not a product but a process of self-discovery through writing; drives him towards it but not far enough to reach it. Despite the vicissitudes suffered by the self in *La Corde raide*, the reader has the sense of encountering a strong personality, troubled, but hectoringly loquacious and with many fixed opinions on life, death, religion, art and human nature. This solid, philosophically coherent self seems to be given in advance rather than to be discovered and take shape in the course of the writing.

La Corde raide then shows the beginnings of an autobiographical project, unsatisfying in itself, groping towards something new. In reaction, Simon then returns to fiction. *Gulliver* (1952), *Le Sacre du printemps* (1954) and *Le Vent* (1957) are all novels of conventional stamp in that whatever they draw from the life of the author is transformed into characters and fictions. Autobiographical material is distanced; formally and thematically it is shaped and trimmed to take its place in fictional patterns. For example, the account of gun-running to Spain in *Le Sacre du printemps* draws heavily on an episode in Simon's life. Much of it is based on personal recollection and some of it, for example the quotations from newspaper, could conceivably be true in the sense that, by scouring the archives of local newspapers on the Mediterranean coast, a meticulous Simon-crazed researcher might be able to verify their existence outside the novel. But the whole episode is organised to play its part in the economy of *Le Sacre du printemps*: this story of disillusioning apprenticeship to life, in the trappings of the 1930s, balances a second story of the same kind set in the urban, apolitical climate of certain bourgeouis milieu in the 1950s; the character plays a double role: naïve youth destined to lose his innocence, mature man called on to preside over the initiation of the next generation.

With *L'Herbe* in 1958, Simon's work begins to move once again towards autobiography. From then on some of his novels share distinctive signs, which, on the whole, become more apparent in each of the novels in which they appear.

In the first place, from *L'Herbe* onwards Simon periodically makes use of material which has a verifiable existence outwith the novels, material which could be seen, read, and touched. In *L'Herbe* the contents of the tin of boiled sweets and Marie's account-books; in *La Route des Flandres* the portraits of the ancestor and his young wife and the contents of 'une de ses malles poilues qu'on trouve encore dans les greniers' (RF, 54);[26] the postcards in *Histoire*; the documentation from the hidden cupboard in *Les Géorgiques*. In each case these deposits left by history – objects, images, texts – sum up an individual life in a peculiar way. Even before she opens Marie's notebooks, Louise knows what she is not going to find:

> elle allait n'y trouver ni journal, ni mémoires, ni lettres jaunies, ni quoi que ce soit de ce genre [. . .] car c'était là des sortes d'idées (tenir un journal, écrire l'histoire de sa propre vie) qui n'étaient même pas capables d'effleurer l'esprit de celle qui les avait tenues. (L'H, 120)[27]

If 'writing the history of your own life' is impossible, as Simon had already discovered in *La Corde raide*, then writing someone else's must surely seem all the more unrealistic. Yet this is precisely the task Louise sets herself in talking about Marie in *L'Herbe*. The function of Marie's treasure is to incite the biographical impulse, to give a base in reality for an impossible project. In *L'Herbe*, as in later novels, these traces of the past are heterogeneous and fragmentary; they can be listed – and no doubt doctored – but it is impossible to exhaust them, to make of them something finished, definitively coherent. They are the grit round which the novels accumulate. They seem to bear authentic witness to the past, yet proclaim the impossibility of reconstructing a life in its entirety.

Where he uses such material, Simon tends to give it an increasingly larger place in his novels. Marie's treasure and the family group photograph occupy relatively few pages in *L'Herbe*, the portraits and the centaur text an equivalent space in *La Route des Flandres*. Postcards are quoted and described throughout *Histoire*; the writings of L.S.M. pervade *Les Géorgiques*. On the other hand, this material serves less and less to construct conventional fictions. In *L'Herbe* Marie's treasure plays an important part in the hazy yet perceptible plot: it binds Louise to Marie and helps explain why she does not leave with her lover. When Louise ceremoniously receives it from the night nurse, she reels under the blow: 'Et, un peu plus tard, quand elle eut ouvert la boîte, restant là à regarder son contenu, sans y toucher, éprouvant toujours cette même perplexité, ce même effarement, les sourcis froncés, silencieuse, et se tenant ainsi, un quart d'heure peut-être, immobile' (L'H, 116).[28] Frowning, silent and motionless: these are traces, indeed even clichés of conventional characterisation. In *Histoire* the fictional is less prominent. Conventional mystery and suspense attend the story of the narrator's relationships with women, wrapped in further enigma by repetition in the life of Oncle Charles. The postcards, however, though providing a skeletal key with which the narrator tries to unlock secrets of the relationship between father and mother, stand for mysteries which are not treated primarily in terms of story or plot. They evoke people, times and places past. Their brief conventional messages, their clichéd photographs, their stamps and date-lines, prove that the past existed, give clues to the values of a vanished civilisation, provoke speculation and interpretation; at the same time they are too elliptical, disparate, coded to give access to the truths at which they hint.

Does fiction make a comeback in *Les Géorgiques*? The case is arguable since discovery of the documents reveals, after 200 pages of suspense, what the old woman had tried to keep secret: L.S.M., the regicide, was also responsible for his brother's death. What could more resemble fiction than this accumulation of dramatic circumstances? 'You have read too many books', said Blum, pouring scorn on Georges's taste for romantic glorification of the family past. But in *Les Géorgiques*, belatedly, Georges gets his revenge, because reality turns out to be as strange as fiction. We know from interviews with Simon and from articles in encyclopedias that the documents Simon uses concern people who existed and events which happened: General Lacombe Saint-Michel voted both for the death of Louis XVI and for legislation instituting the death penalty for royalists who took arms against the Republic; consequently, his brother was condemned to death. As in *Histoire*, then, Simon is embroidering and inventing not on the basis of 'nothing at all', as Blum puts it, but on documented, historical facts. Thus in using documents and by the way he uses them, Simon moves his novels away from the purely fictional towards the domain of history and historical biography. He tempts the reader to apply to his novels a criterion alien to fiction but central to historical study and frequently evoked in studies of biography and autobiography: to what extent can the truth of what is advanced be verified by appealing to sources external to the text?

Furthermore, biography slides toward autobiography in that the principal characters belong to a single family. Through time that family changes shape to resemble more and more Simon's own family-tree. The family is constructed in two stages. To begin with Simon creates a fictitious family modelled on the paternal side of his own family but incorporating other elements. The Thomas family appears for the first time in *L'Herbe*. To a large extent, the life and death of Marie reflect those of one of Simon's paternal aunts. Marie's tranquil courage, her self-sacrificial devotion to her younger brother, her flight to the south-west in 1940, the sale of the family house: all this is taken from life. But if one compares the Thomas family with Simon's family one can see the extent to which Simon has adapted it to the purposes of his novel. (See page 100.)

The form of the family tree is recognisably the same. *L'Herbe* even contains a passing reference to the profession of two of the sisters: one entry in the account-book mentions 'our two salaries' (L'H, 121). But the oldest sister Louise disappears, bequeathing her name to an invented character, an incomer to the Thomas family, Georges's wife. Antoine, the career officer, becomes Pierre, professor of philology. The portrayal of Sabine may owe much less to Simon's mother than to his grandmother.

In *La Route des Flandres* Simon, the traditional novelist, is still at work in the construction of character. The Thomas family survives and even, in the passing, expands through the single mention of two sisters for Georges, Christine and Irène (markers for a potentional development which came to nothing?) An important new member of the family is added on the maternal, Sabine's side. To produce this figure Simon reworks events and personalities from two sources: an officer killed by sniper fire in Flanders in May 1940 shortly after his squadron had been massacred; the old uncle in *La Corde raide*. From this fusion springs Captain de Reixach, Georges's distant cousin. His outward circumstances are those of Colonel Rey; his respect for good form, for a military, aristocratic code of behaviour and speech relate him to the old soldier of *La Corde raide* who, as death approaches, sustains himself by paying 'meticulous attention to appearances' (CR, 25). De Reixach's inner emptiness comes from neither of these sources. Simon invents the imbroglio of de Reixach's relationships with Corinne and Iglésia, and the sexual fantasies of Georges and Blum.

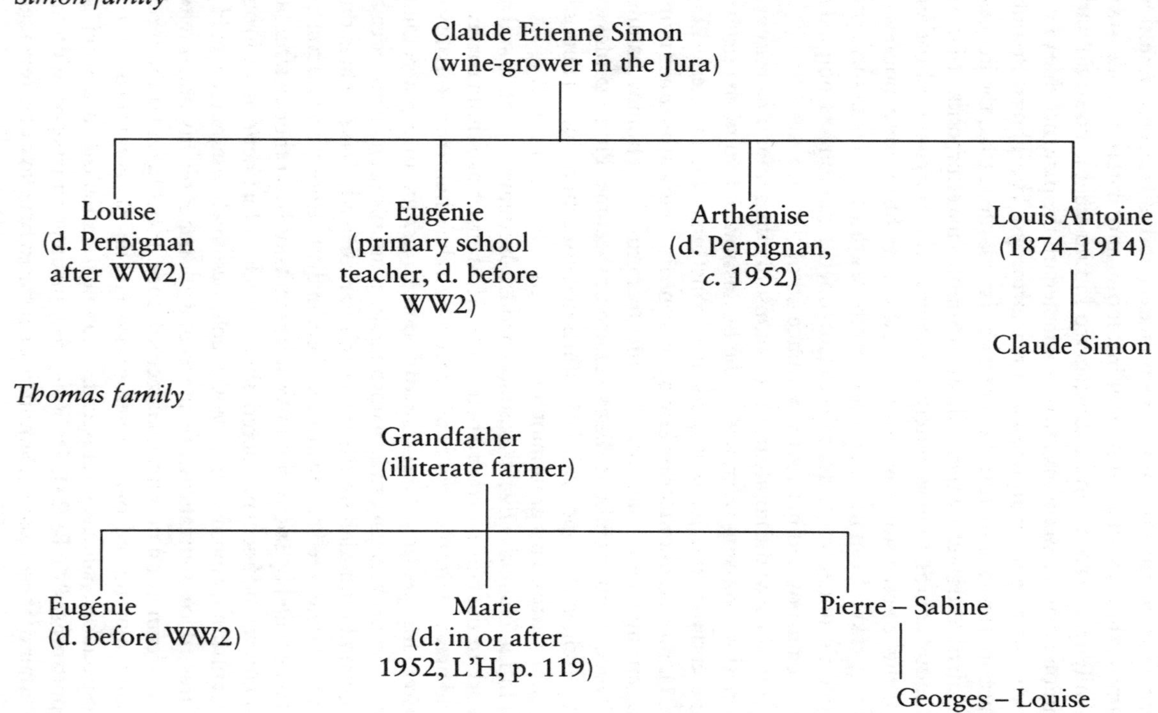
Simon family
Claude Etienne Simon (wine-grower in the Jura)
Louise (d. Perpignan after WW2)
Eugénie (primary school teacher, d. before WW2)
Arthémise (d. Perpignan, c. 1952)
Louis Antoine (1874–1914)
Claude Simon
Thomas family
Grandfather (illiterate farmer)
Eugénie (d. before WW2)
Marie (d. in or after 1952, L'H, p. 119)
Pierre – Sabine
Georges – Louise

These revisions of the family in *La Route des Flandres* do not in themselves threaten to break the referential illusion. Simon dispenses with one character (Louise), introduces others (de Reixach, Corinne, the ancestor general) and extends the range of interests of existing characters (Sabine and Georges are fascinated by lineage). Such practices were standard in the nineteenth-century novel. Balzac frequently dropped, altered and rearranged his characters. Readers were no less inclined to believe in them, no doubt largely because readers recognise that the illusion of reference is precisely that, an illusion which they willingly accept for the purpose of entering imaginatively into other people's lives. Part of the pleasure lies in knowing that events and characters are not real: Georges is a fiction so I am not exposed to the perils of the Flanders road but safely tucked up in bed with a book. But part of the pleasure also lies in fantasising with and about Georges. *La Route des Flandres* is not a realist novel in that it lays bare the idealism which subtends realism and ultimately deprives the reader of the certainties of knowledge which Balzac claimed to provide. But everything which undermines representation in *La Route des Flandres* must be set in the balance against all that makes readers believe in Georges, Blum, Iglésia, de Reixach and Corinne. The overwhelming effect of the novel depends on the reader's perception of Simon's extraordinary balancing act in creating illusions which he simultaneously destroys.

In the course of a second phase, the members of the Thomas family gradually disappear, to be replaced by new characters who issue more directly from Simon's own family. *Histoire* marks a decisive step. The Thomas family vanishes, for a simple reason: the parent figures, Pierre and Sabine, are incompatible with names, dates and information on the postcards. In place of these parents, Simon constructs speculative, unfinished portraits of two other characters: his own parents. The father, an officer in a regiment of 'colonial infantry', serves his time in the colonies, including Madagascar. After a long courtship he marries a young Frenchwoman who accompanies him on his final tour of duty. He is killed at the Front near Luzy in the Department of Meuse. The mother comes from a respectable, Catholic, propertied family established in a town in south-west France. She venerates her husband's memory. With her mother and one of the boy's uncles, she begins the task of bringing up her only son. She dies after a long and painful illness when her son is still at a tender age. These remarks apply both to the new characters in *Histoire* and largely to Claude Simon's family circumstances.

In *Les Géorgiques* Simon revives and revises the family from *Histoire* in that of 'the boy . . . become a man' (LesG, 215). Mother and grandmother are both present. So too are Corinne and Uncle Charles, even if their roles are cut back, thus reducing the elements of pure fiction. In a curious phrase Simon speaks of 'the series of bereavements which had struck the family' (LesG, 182), curious because only one recent bereavement – the mother's death – is mentioned in *Les Géorgiques*. What other deaths belong to this series? Readers of *Histoire* could supply one: the death of the boy's father. But not until *L'Acacia* will Simon add a third, the boy's uncle, poet and Deputy, who also served and died in the Great War. Like 'the two salaries' mentioned in *L'Herbe*, then, this is an example of the interpenetration of fictional and historical worlds, all the more striking in that it occurs here in a fragment of text composed by Simon, not, as in *L'Herbe*, in an incorporated historical document. In this case, the historical has not yet been perfectly rewritten into Simon's family fiction.

Above all, *Les Géorgiques* corrects one aspect of the family which was thrown out of joint by *Histoire*. Although abandoning the Thomas family in *Histoire*, Simon clearly wished to preserve a suggestion of continuity with *La Route des Flandres*. He therefore retained Corinne and her marriage to de Reixach. But though she still remains a focus for sexual fantasy, Corinne's relationship to the narrator is altered. She becomes his youthful first cousin, brought up in the same house. The effect is to expel de Reixach from the narrator's maternal family and consequently to detach that family from its ancestral past. *Les Géorgiques* restores the family's lineage. This is no simple task. Simon's mother's family is linked by descent to two generals, Jean-Pierre Lacombe Saint-Michel and General de la Houlière. Lacombe Saint-Michel was a member of the Convention; de la Houlière committed suicide after the French army under his command was defeated by Spanish troops in the Roussillon campaign of 1793.[29] In *La Route des Flandres* Simon conflates these two figures, and exploits in addition the portraits of de la Houlière's son-in-law, Lemoine Daubermesnil, and his wife, Thérèse Vaquer, portraits which hung in Simon's family home in Perpignan. In *Les Géorgiques* Simon concentrates on the Lacombe Saint-Michel side of the family and establishes a line of descent from L.S.M., to renegade son, to false baptist pastor, to grandmother, mother and finally to the 'boy/visitor'. These stages correspond to the generations in Simon's own family. As a result, the name and lineage of Reixach is expunged: the cavalry officer who led his men into an ambush in Flanders becomes an anonymous major. In 1961 Simon remarked to an interviewer that he had retained enough stories about his family to fill his books for a long time.[30] In *Les Géorgiques* in 1981, he at last reached the point where his family become the basis for his fiction.

This, then, is one trend in Simon's work. Increasingly, he does not invent fictions but bases his novels on historical documents and stories about his own family. In these circumstances, the reader cannot but begin to see the recurring narrator–protagonist as something of an autobiographical figure. Simon does not discourage this view. Twice in *Les Géorgiques* he invites the reader to reflect on the autobiographical dimension of his fiction by formalising links between this and previous novels. First, while describing the débâcle of 1940, he builds a bridge to *La Route de Flandres*:

> Environ deux heures plus tard, il chevauche de nouveau à côté d'un autre cavalier derrière le chef de l'escadron et un lieutenant (il rapporte dans un roman les circonstances et la façon dont les choses se sont déroulées entre-temps: en tenant compte de l'affaiblissement de ses facultés de perception dû à la fatigue, au manque de sommeil, au bruit et au danger, des inévitables lacunes et déformations de la mémoire, on peut considérer ce récit comme une relation des faits aussi fidèle que possible: le carrefour et les champs parsemés de corps, le blessé ensanglanté, le mort étalé au revers du fossé, sa progressive reprise de conscience, sa brusque décision, sa course haletante en remontant la colline dans les prés coupés de haies d'aubépine, le franchissement de la route où patrouillent les auto-mitrailleuses ennemies, sa marche dans la forêt (Et où irez-vous?), sa soif, le silence du sous-bois, le chant du coucou, les bruits lointains de bombardements, la rencontre imprévue des deux officiers rescapés de l'embuscade, l'ordre négligent qu'il reçoit de monter sur l'un des deux chevaux de main conduits par l'ordonnance, la traversée de la ville bombardée, etc.). (LesG, 51–2)[31]

This delightfully teasing passage proclaims both the autobiographical intention and the impossibility of autobiography. How faithful to the facts can an account be when it is 'as faithful as possible'? To judge by all the hints given here, the answer must be somewhere between 'not very' and 'not at all'. Perception and memory are unreliable. Paradoxically, the narrator of the previous 'true account' was a character in a novel ('il raconte dans un roman'), just like the 'il' whose similar adventures are recounted in *Les Géorgiques*. Yet the adventures, according to this paragraph at least, are not just similar but the same, a point repeatedly hammered home by use of the possessive pronoun and the definite article. So the autobiographical criterion of truth to the facts seems to be respected in the mutually reinforcing coherence of the two accounts.

Coherence is also affirmed in another passage. Three successive paragraphs describe associations evoked by the idea of death:

> *longtemps ainsi l'idée de la mort restera associée pour lui au parfum de l'eau de Cologne imbibant ces éponges qu'on leur mettait dans la main lorsque enfants, pendant la Semaine sainte, habillés en pénitents, on les conduisait auprès de ces Christs décharnés exposés sur des lits de fleurs et dont ils essuyaient les orteils avant d'y déposer leurs lèvres.*
>
> *Plus tard l'idée de mort se confondra pour lui avec l'odeur écoeurante d'huile chaude et rance qui imprégnait la nourriture servie aux volontaires étrangers dans la grande salle à manger de ce palace de Barcelone réquisitionné.* [. . .]
>
> *Plus tard encore et pour de longues années, cette même idée de mort sera pour lui inséparable des noms d'une suite de hameaux ou de villages s'échelonnant entre la Meuse et la Sambre, déserts dans la campagne déserte secouée de loin en loin par les échos des explosions.* (LesG, 225–7)[32]

Here the experiences of a pious childhood, of Barcelona during the Civil War, and of Flanders in 1940 are explicitly ascribed to the memory of a single individual. This passage is the equivalent, in Simon's fiction, of the lines quoted above from *Le Discours de Stockholm*: it affirms the continuity of the self through time (though the self refrains from speaking in the first person). So Simon's invitation to re-read his novels differently, autobiographically, is not limited to the matter of 1940 (told and retold in *La Corde raide*, *L'Herbe*, *La Route des Flandres*, *Histoire*, *La Bataille de Pharsale*, *Leçon de choses* and *Les Géorgiques*) but extends to Spain in 1936 (*La Corde raide*, *Le Sacre du printemps*, *Le Palace*, *Histoire* and *Les Géorgiques*) and stretches back to childhood (*Le Tricheur*, *La Route des Flandres*, *Histoire*, *La Bataille de Pharsale* and *Les Géorgiques*). Thus Simon reclaims his work, transforms it retrospectively into a form of autobiography, declares engagingly that he has spoken about himself – in so far as such a thing is possible.

What then is the nature of this autobiographical project? How does it affect our reading of the novels? In so far as *Les Géorgiques* is concerned, Simon is reacting against the critical orthodoxy of the later 1960s and 1970s. He is reasserting, against Ricardou, the proprietorial rights of the author. In *Les Géorgiques*, for the first time, Simon not merely re-uses previous episodes from his fiction but comments on their re-use. Like Sarraute in *Enfance*, he grounds them in personal experience and links them to a single continuing self. The effect, at least for some readers, is to add a dimension of pleasure to the reading of *Les Géorgiques* and earlier novels, which critical authority had previously denied. The Claude Simons of *Les Géorgiques* are legion. Of the three new versions of Georges, the cavalryman's experience comes closest to that of the historical Simon. It seems possible that in Chapter 2, Simon is reworking a version of the events of 1939–40 originally written for possible inclusion in *La Route des Flandres*; but in the débâcle of 1940 order collapsed more completely than in the manoeuvres and delays of the phoney war; so in *La Route des Flandres* winter episodes had to give way to those of spring. O. blends another Georges, George Orwell, with Simon's previous galaxy of innocents abroad in Spain. The O. section, aș distinct from the account of the winter war, shares a characteristic common to Simon's Spanish narratives in *Le Sacre du printemps* and *Le Palace*: the experience of a young man is seen from the disillusioned perspective of an older narrator: two Claude Simons separated by time. The focus of the ancestor sections in more variable still. An old man, the one who holds the pen and has written previous novels, recounts what 'the visitor' saw and reconstructs the childhood perspective of a young boy: at least three Claude Simons already. But equally – I shall return to this theme in Chapter 7 – the ancestor is also a fictional Simon: soldier in his youth, writer of memoirs in his old age. ' "*[C]royez-vous que j'aie tant d'années à jeter par les fenêtres?*" ' These last words of *Les Géorgiques* present an old man in a hurry. L.S.M.'s urge to build, to construct, endures even in the rambling incoherence of old age. Surely a wryly distanced image of the writer whose masterfully ordered novel we have just finished reading.

But the autobiographical project in *Les Géorgiques* is in no sense a sudden lurch, a dance in tune with the later 1970s and early 1980s. Its origins go much further back. The effect of *Les Géorgiques* is to set the history of Simon's fiction in a new perspective. The use of material fragments of the past, the rewriting of family history, become more central; *L'Herbe*, *La Route des Flandres*, *Histoire*, *Les Géorgiques* become the backbone of Simon's work; the novels from *La Bataille de Pharsale* to *Leçon de choses*, an appendix (though I should not wish to draw all the possible conclusions from this medical metaphor). Despite this shift, Simon's autobiographical fictions remain firmly within the genre of the novel. Formally, they propose no autobiographical pacts: the shadow of Claude Simon flits between the many versions of Georges. More important still, each of these characters has their separate existence. O., L.S.M., the cavalryman, Oncle Charles, all in their own ways are not Georges, far less Claude Simon. Simon's subject is far from exclusively himself. The spotlight is not on the inner experience of an individual or on the development of a personality. In the long run Tolstoy rather than Dostoevski is the appropriate comparison, although a Tolstoy who has read Proust and Faulkner. Simon's scope includes the self, family, society and history.

In so far as the novels are indeed autobiographical, they produce autobiography of a distinctive kind. Simon's growing desire to base his novels on his own and his family's past goes hand in hand with the continuing conviction that the past cannot be successfully reconstructed. The aim of much autobiographical writing is to give a definitive version of the past. Chateaubriand proclaims it in his title: *Mémoires d'outre-tombe*; in *Les Mots* Sartre mocks it in his youthful dream of being a 'posthumous' writer.[33] For Simon, as for Sartre, this aim is an illusion. In many interviews Simon has quoted the example of Stendhal describing how he crosssed the St Bernard Pass on Napoleon's march into Italy:

> Soudain, tandis qu'il écrit, il se rend compte qu'il est en train de décrire non pas ce qu'il a vécu mais, dit-il, une gravure représentant cet événement, gravure qu'il a vu cinq ou six ans après et qui (ce sont les termes qu'il emploie) a, depuis, 'pris la place de la réalité'.[34]

For Simon that engraving represents all that has been read, seen, experienced by the writer since the event, whatever it may be, took place. Perhaps above all, the writer's experience includes what he himself has written. Simon's successive rewritings of the past from novel to novel transform it and take the place of what went before. As Anthony Pugh has put it, what Simon produces is 'a continually revised *alter*-biography in which the ideal of self-knowledge is permanently deferred'.[35]

And yet the pursuit continues. It has a focus: the search for origins. Is it possible to make sense of this search by approaching the novels psychobiographically?

6

Psychobiographical fictions: *Le Tricheur* to *Leçon de choses*

Simon's novels are often about the family, about identity and its origins. They show the weight of the past on the present; they explore the interplay between language and memory; their construction has come more and more to depend on the play of association and analogy. While all this might suggest that they cry out for psychocritical interpretation – indeed, Lucien Dällenbach has gone so far as to say that they imply it and involve the reader in it[1] – that cry went for long unheeded. One important reason for this was certainly the direction taken both by Simon's work and Simon criticism from the mid-1960s to the late 1970s. As we have seen, the phase of Simon's work of which the beginnings can be seen in *Histoire* (1967) and which strengthened and developed as far as *Triptyque* and *Leçon de choses* was closely related to the adventure of structuralism, and more particularly to the version of structuralism described and prescribed by Jean Ricardou. However, criticism informed by psychoanalytic theory slowly gained ground in the 1970s, and in the 1980s it came into its own.[2]

No doubt *Les Géorgiques* encouraged critics to re-read Simon in a new perspective: the return of story, history, character, family and autobiography raised questions which Ricardolian criticism was hard put to answer. But the reaction against Ricardou was already under way, and it was primarily this critical context, aided by the new shift in Simon's work, which explains the characteristics of the psychoanalytic revolution. In the first place, psychoanalytic criticism began to reconsider what Ricardou had suppressed: the representational elements in Simon's work. In doing so it has often stressed what equally went missing: the humanity and universality of his writing. Sometimes it has done so directly: Anthony Pugh emphasised how *Histoire* deals with the shared human experience of the death of loved ones.[3] More often critics have drawn implicitly or explicitly on Freud's view of the nature of the work of art and how it appeals to its reader or spectator. Representations in a work of art have to do with fantasy and desire; through the devices of form such fantasies are made accessible and acceptable to others.[4] So, to some extent a new Simon emerged, or at least a revised version of an old Simon: the critical emphasis was not on linguistic play as formal experiment, but on the shifting relationships between linguistic exploration, the search for origins and the construction of identity.

Second, it has been characteristic of psychoanalytic criticism to draw upon a varied but limited range of psychoanalytic thinkers: Freud, of course, and to some extent his standard mediators, Laplanche and Pontalis; Marthe Robert, in particular her elaboration of Freud's 'family romance' in its literary expressions; above all, Lacan. Celia Britton in particular has found in Lacan a wealth of ideas to illuminate Simon's texts, from the central thesis of her book – the relationship between vision and desire – to specific manifestations of that relationship related to the notion of the mirror phase. The central appeal and aptness of Lacan, however, has lain in the parallel between this psychoanalytic theorist and Simon: Lacan, like Simon, although more than a decade earlier, took the so-called linguistic turn. That is to say that Lacan's reinterpretation of Freud is largely based on the deployment of structuralist linguistics. The truth of the unconscious does not lie behind words, but in words, in meaning which is constantly glimpsed and deferred as it slips along the chain of signifiers. Metaphor and metonymy are its modes of movement. Behind Lacan, as behind Ricardou, lies a version of Saussure.[6] This common ancestry has given Lacan purchase on Simon, and occasionally vice versa.[7]

It follows partly from this that a third characteristic of psychoanalytic criticism has been caution. Lacan is opposed to psychoanalysts who study varieties of object relation or attempt to chart stages in the development of human personality. This distinguishes him both from Freud himself and from many of those who have most fruitfully and persuasively elaborated his theories, for example, Melanie Klein.[8] By contrast, Lacan detriangularises the Oedipal drama; for him the essential phase in human development is access to the symbolic order of language, and this is a drama which ends only with life itself. Psychoanalytic criticism of Simon, then, has tended to shy away from questions of development – and even more from any attempt to offer a comprehensive unifying account of Simon's life and work. Such a task may one day be attempted, using charts supplied by a different range of psychoanalytic geographers. But although Simon's life provides dramatic material for such an endeavour, critics using psychoanalytic concepts have not used this material. Many such critics share the widespread perception that such criticism simplifies and falsifies. It is also too soon. Too little is known about Simon's life to make any such attempt credible.

My aim similarly is to look at the novels, not the novelist. The use of psychoanalytic theory in this chapter will be eclectic and partial, since its main function is not to make the novels fit into a single explicatory scheme but rather to see whether psychoanalytic thought can elucidate patterns in the work, like a magnet revealing lines of force. In particular, though not exclusively, I will use a pre-Lacanian Freud, because of the question I want to address. One of the most striking characteristics of Simon's work is that it has proceeded backwards and forwards: the novels grow out of one another, but in doing so they rework and transform the same material: motifs, themes, characters, aspects of form. Now it is a commonplace of psychoanalytic theory that the unconscious also works in this way, digesting the present, reshaping and transforming the past. If then we take seriously Freud's assertion that the literary work is a daydream in which fantasies are given shape, should we not be able to follow in the succession of Simon's works the traces of that psychic reshaping? Can psychoanalytic theory throw some light on the development of the novels from the 1940s to the 1970s?

The constantly repeated and defeated aim of Simon's novels from *Le Tricheur* to *Histoire*, explicit from *Le Vent* onwards, is to reconstruct the past. One focus of that aim concerns family relationships. Anthony Pugh picked out a phrase from *Le Sacre du printemps* to remark that 'the central place that is occupied by memory in Simon's novels during the first and second "periods" is directly related to a paradox – "la présence de cette absence" '.[9] Lucien Dällenbach spelt it out: 'le roman simonien, essentiellement matrilinéaire, est bâti autour d'un vide peuplé par une absence essentielle, celle du Père'.[10] Sometimes the father is literally absent. *Le Tricheur* and *Le Sacre du printemps* foreshadow *Histoire* in that the father has been killed in war, leaving behind the vaguest and most uncertain of memories. the originality of *Le Vent* is to propose a variant and a double image of response to the loss of the father. Montès returns to the town of his birth in order to take the place of the father he had lost as a child. He fails – just as the narrator struggles to reconstruct the truth about a model figure, Montès, whom he perceives to be both naïve and heroic. Similarly in other novels absence becomes presence in the guise of one or several substitute fathers. In *Le Tricheur* the role of father to Louis is distributed between three characters: his guardian, Armand and the priest. In the two parallel stories of *Le Sacre du printemps* the stepfather is to Bernard as Ceccaldi is to the stepfather. In both these novels Simon uses Freudian patterns conventionally arranged. *Le Tricheur* shows an unsuccesful attempt to surmount the Oedipal complex. Louis fulfils the male child's fantasy of killing the father; but his murder of the priest leaves him only with a sense of emptiness, unable to take that father's place and assume an active role in the world. By contrast, in *Le Sacre du printemps*, as the title suggests, two Oedipal conflicts are successfully resolved: each of the young men, by different routes, grows into renouncing absolutes and accepting primordial disorder; Bernard acknowledges his stepfather's fraternal welcome into the shared disabused world of adulthood.

In *La Route des Flandres* such father–son relationships are woven in a more complex way. *La Route des Flandres* dramatises death in the context of war: the living perish, the material world dissolves, there is, beyond appearances, no ultimate secret to be unlocked except that 'décevant secret qu'est l'absence de tout secret et de tout mystère' (RF, 270). Three father figures resist this universal dissolution: Pierre, Georges's real father; de Reixach, his cavalry captain; and the Revolutionary general who is the common ancestor to Georges and de Reixach. Each of these figures provides Georges with a kind of model. Dällenbach has spoken of Simon's repeated creation of 'une image paternelle injonctive, avec qui j'entretiens un rapport d'admiration et d'émulation'.[11] Certainly there is an element of admiration in the portrayal of the ancestor and de Reixach. Both these figures may be linked with the family romance: the child substitutes for the real father an imaginary equivalent who is high-born and prestigious.[12] In particular Georges clings fiercely to the ancestor as a figure of heroic proportions, of glorious battle and tragic death.

Yet, much more than the prestige, it is the ultimate inadequacy of each of these models which is most striking. The ancestor's suicide may be neither the fruit of intellectual disillusionment nor of a heroic refusal to outlive the shame of defeat in battle, but rather – what is presented as much more mundane and sordid – a reaction to the discovery of his wife's unfaithfulness. Pierre, the humanist intellectual, stands for the attempt to understand history, to see it as order and progress. His lament for the destruction of the library at Leipzig is countered by Georges's protest that if all the books stored in that library were insufficient to prevent its destruction, what purpose did they serve? Pierre embodies one of Simon's constantly recurring themes: the denigration of writing and writers who stand accused of embellishing and falsifying reality.

In the relationship between Georges and de Reixach, Simon most overtly writes a further revised version of the Oedipal drama. Reluctant to have anything to do with this distant relative, Georges finds himself compelled into admiration and dependence: de Reixach's straight back is the one fixed point in the dissolving world. Yet all Georges's fantasies aim to wrench de Reixach from his pedestal, to prove that his apparent self-control and insouciance mask the inner emptiness of a life destroyed by military incompetence and his wife's infidelity. In *La Route des Flandres* part of the Oedipal fantasy is enacted. In making love to Corinne, his captain's widow, Georges takes the father's place – but in vain: he gets no closer to him, no nearer to finding the truth about him and about Corinne, which the novel presents as yet another possible key to all truth. The truth here would be certain knowledge of the primal scene, the sexual encounter to which Georges could definitively trace his origin. But just as he cannot find his way back to that, so he cannot go forward. As in *Le Tricheur*, the novel leaves the Oedipal crisis unresolved. A single image encapsulates the mingled intense emotions, ranging from adulation to derision, which Georges attaches to de Reixach: sabre drawn and raised to avert a sniper's bullet, de Reixach crashes to the ground. But this destruction of the father's phallic power is never complete. The image obsessively recurs. As often as he falls, de Reixach rises again. Iglésia refuses to verify his death; 'l'ai-je vraiment vu ou cru le voir?' (RF, 314); Georges's hunger for that death is never satisfied. De Reixach, Pierre and the ancestor represent possible model identities, none of which can be satisfactorily assumed. Consequently the identity of Georges remains uncertain and unformed, wandering between poles which provide no secure resting place.

In recent years, critics have commented, not without disapproval, on the scarcity of sympathetic female characters in Simon's work. Women in his novels, it has been claimed, are irremediably other, conventional male stereotypes, whether as sex objects or objects of suspicion.[13] Celia Britton, while endorsing such views in general, remarks that she is not troubled by this characteristic since female figures in Simon are so clearly the projections of unconscious and semi-conscious desires and fantasies.[14] This generous response may not entirely suffice to restore Simon's universality. If it is the case that the pleasure of the text comes from shared fantasies, and Simon's fantasies are exclusively or largely male, then much of his work must remain inaccessible to female or some female readers – at least as long as one assumes that male and female fantasies are the distinct and exclusive possessions of those who are biologically male or female, which Freud would deny. Jean Duffy entitled her article in this field '(Ms)reading Claude Simon: a partial analysis', thus inviting critics to a fuller or at least other partial readings. One might remark that the apparent fault lines between men and women in Simon's fiction are not rigid. In general, the early novels give women an ambiguously privileged, albeit clichéd place as beings who live close to nature and in harmony with it. Belle and her mother in *Le Tricheur*, Josie in *Le Sacre du printemps*, and the line extends to Marie in *L'Herbe*, all 'acceptent la succession des sensations sans chercher à leur découvrir un lien' (LT, 27). But women are not alone in this. The De Chavannes brothers in *Gulliver*, Ceccaldi in *Le Sacre du printemps*, the grandfather in *L'Herbe* and the peasant soldiers of *La Route des Flandres* are all endowed with the prestige of a natural integrity. And on the other hand, the long line of Simon's lucid, tormented figures, though predominantly male, includes not just Louise in *L'Herbe*, but also, in their own ways, Eliane in *Gulliver*, Edith in *Le Sacre du printemps* and Cécile in *Le Vent*.

Nevertheless, in the context of a psychoanalytic reading, it may be most appropriate to expand Britton's argument and her implied plea for understanding. Women in Simon's fiction bear a heavy emotional charge. They play many parts and, in particular, their sexual role is never far divorced from their role as mother. In *Le Vent*, for example, although Montès sets out to find his father, the drama is played out almost exclusively in relation to his mother. Three women – Hélène, Cécile and Rose – embody different fantasies of women's relationships to their bodies, to sexuality and to childbearing. Rose engages the most complex emotional response. She is prospective lover, openly affirming and accepting her sexuality. But it is in her role as mother that she attracts Montès. This is strongly emphasised in the imagery which describes her; it explains both the fascination she exercises on Montès and the timidity of his approach. Attempting to identify with the father, he has come to the necessary rite of passage which is desire for the mother. Rather than moving forward from this point, however, the story moves backwards. The identification between Rose and Montès's mother is heightened by Rose's death, which reactivates the grief he felt at his mother's loss. And that grief in turn comes to bear a strong pre-Oedipal charge:

> [C]omme si, assis là dans le temps aboli à côté de Rose morte, enfermé, enfoui dans cette chair, ce lourd parfum de lilas en train de se fâner, de se flétrir lentement, il se trouvait ramené à un état en quelque sorte foetal, lové dans la douloureuse et torturante (dit-on) quiétude d'une vie intra-utérine dont il allait être – pour la seconde fois, et pour la seconde fois d'entre les cuisses d'une femme, bien que celle-ci fût de cinq ans plus jeune que lui – expulsé, projeté, hurlant et terrifié dans le vide. (V, 186)[15]

Grieving for these two deaths turns to grief over a lost one-ness and over the initial expulsion into time.

Dällenbach's claim that Simon's novels are essentially matrilinear is backed up by Britton who points out that, except in the case of the mother's second husband in *Le Sacre du printemps*, substitute father figures come exclusively from the mother's side of the family. 'It is as though Simon's protagonists remain in some sense tied to the imaginary dual relationship with the mother: even the father figure has strong maternal associations – in, in fact, a sort of male mother figure.'[16] Repeatedly the mother is discovered behind the father, not as a mere delegate transmitting the father's values, but as the creator of father and values. In *Le Tricheur* it is Louis's mother who proposes two possible model roles for him: soldier and priest.[17] In *Le Vent* Montès has been wrested as an infant from his father's culture and values and shaped solely by his mother. In *La Route des Flandres* Sabine creates the legend of the family and introduces Georges to de Reixach through the letter. The suspicion and fear which often mingle with fascination in Simon's characterisations of women may partly be explained by this: unconscious desires and fears are projected predominantly on the female sex. Fantasies concerning father figures have female associations; mother figures play male roles. Thus in *Le Tricheur*, although Louis kills a priest, the murder scene emphasises overtly female attributes of the priest's speech and dress, and Simon colours the scene still further by juxtaposing images of mother with those of priest. Conversely, in *Le Vent* Rose is predominantly a mother figure: Montès's timid love for her illustrates with extreme delicacy the classic Oedipal desire to take the father's place. Yet Montès is also indirectly responsible for Rose's death, so that in this relationship one may also discern a faint shadow of the unconscious desire to kill the father, with its accompanying pangs of grief and guilt. Similarly, the pre-Oedipal role of mother has a troubling ambiguity in the novels. Rarely do Simon's male protagonists long unambiguously for the return to a perfect harmony with the mother. Why should the quietude of the womb be 'douloureuse et torturante' (V, 186)? Perhaps because women in Simon also play the father's role, in that as givers of the law they threaten castration. Hence the recurrent figure of the old crone whose sharp knife gleams so menacingly in *Triptyque* (quoted and commented on in Chapter 3). That Oedipal fear casts its shadow over the pre-Oedipal idyll.

Of the novels I am considering here, *Histoire* constructs family relationships most subtly. It has a different tone from previous novels. Less anguished than *Le Vent* or *L'Herbe*, lacking the desperate excitement of *La Route des Flandres*, it moves at a slower pace, its mood more autumnal. Although the family circumstances of the narrator and protagonist come closer than ever before to resembling those of Simon himself, *Histoire* is much less than previous novels a reconstruction of the past, in so far as that phrase implies the prior existence of a single truth. Words are known to be vain, but words and pictures are all we have and these Simon sets to work, building fragments from fragments, from images in books, photographs, postcards and their laconic messages. If in *La Route des Flandres* Simon's aim seems to be to master the past, to attain knowledge, in *Histoire* he comes closer to accepting exclusion from knowledge, the impossibility of attaining mastery.

To this shift corresponds a different thematic emphasis. *La Route des Flandres* is concerned with power relationships, with confrontations between son and multiple father-figures. *Histoire* is more reminiscent of *Le Vent* in giving pre-eminence to relationships with women: grandmother, model, Hélène, Corinne and the narrator's mother. But *Histoire* reverses the focus of *Le Vent*: in place of a young woman – Rose – seen as mother, *Histoire* imagines the mother as a young woman. A yet greater difference between the novels is that the characters in *Le Vent* are firmly contoured. Rose embodies various fantasies, Oedipal and pre-Oedipal; but however many, however contradictory they may be, she is held within a conventional narrative framework, clearly distinguishable from Cécile and Hélène, even if they too figure some of the same fantasies. In *Histoire* outlines remain distinguishable and emotions attach themselves to specific characters: awe to the mother, the pain of loss and guilt to Hélène, incestual desire to Corinne, desire and the fear of sexuality to the model. Yet the new freedom of Simon's complicity with language in *Histoire* means that more even than in *La Route des Flandres* characters merge and fuse. 'l'une d'elles touchait presque la maison' (H, 9): the first phrase of the novel fittingly foreshadows the importance of the female, close, so close, yet always just out of reach. The pronoun without an antecedent is equally characteristic, one of a number of techniques by which Simon creates ambiguity and uncertainty of reference. He runs scenes into one another, juxtaposes paragraphs not diegetically linked, attributes the same motifs (for example the eating and drinking of chocolate), and indeed the same names, to different characters. By these means, and others, men and especially women melt and flow into one another like the leaves of the acacia tree reflected on the house wall, 'un reseau mouvant de taches et d'ombres entrecroisées se faisant et se défaisant sans trêve' (H, 41);[18] and this allows feelings which are complex, conflicting, more or less avowable, to be displaced and to circulate from one figure to another.

Commenting on this circulation, critics have drawn attention to the importance of Oedipal relationships in *Histoire*. Celia Britton pointed to evidence of unconscious desire for the mother. Linguistic parallels between the first description of the mother and the first sight of the model (H, 18–19, 295) encourage us to read the second scene with the model as 'a slightly more "acceptable" reinvention of what cannot be expressed in the first scene with the mother: a displacement on to the model, the sexually available woman par excellence, of the idea of the mother's sexuality'.[19] Anthony Pugh emphasised the mother's phallic role. 'The image of the mother is treated', he argued, '. . . with an almost fearful deference'. In the imagery of the mother's death-bed scene – the knife-blade, the tonsured, apparently decapitated priest, the father's photograph – he found a symbolic staging of the fear of castration.[20] In the father's absence, the mother wields symbolic power; from her values, 'les croyances maternelles et les leçons de catéchismes' (H, 43),[21] the narrator and one of his doubles, Lambert, seek to free themselves through wordplay.

More important than the Oedipal relationships however are the pre-Oedipal, that is to say, traces of a relationship felt to have existed between mother and child before the shadow of the father fell fully across it.[23] *Histoire* is a novel written from exile affirming the desire to return home to the mother. It moves from night to day, then back to night again. Night favours recollection and reintegration of the past: the first chapter reconstructs a family tree in various senses. Day breaks the umbilical cord – 'mon double encore vacillant au sortir des ténèbres maternelles, fragile, souillé protestant et misérable' (H, 45)[24] – and exposes the narrator to the exile of the present, the dispersal of the family home and of the mother's body, symbolised in her collection of postcards, now untied and scattered. Marthe Robert commented on Don Quixote's rejection of the present and nostalgia for a mythical past:

> la civilisation est maudite en tant qu'oeuvre du père créateur qui a mis l'Histoire en mouvement; elle appartient à la catégorie du mal et de la séparation parce que, ayant imposé au monde la double fatalité de la sexualité et de l'appropriation du sol, elle rompt pour toujours l'union du petit enfant avec 'les pieuses entrailles de notre première mère'.[25]

Histoire shows this same forceful rejection of the fall into history: the narrator laments the disruptive force of sexuality ('nous ne pouvons pas nous perdre'[26] – but we will and we have); he parodies the clichés of those who set out to master history, for example the puffed-up triumphalism of the quotations from John Reed's *Ten Days that Shook the World*; and he caricatures those who seek to appropriate the earth by buying and selling: imagery from classical mythology and from the natural world sets banker, antique-dealer and Paul, the property speculator, in contexts which render them at once threatening, puny and ridiculous.

In the latter part of the novel, the reverse movement from day to night quickens the reader's sense of the origin of this rejection of the present. The narrator drives back towards town and home into the night, 'pensant que je pourrais continuer ainsi avançant toujours m'enfonçant dans ses entrailles tièdes sans rien voir d'autre que les ténèbres rassurantes' (H, 324).[27] Whereas the early part of the novel shows the priest standing between mother and child, the last section records evidence of the father's death: the cemetery in which his body lies, a colonel's note of condolence. The novel ends in a further backward movement through time and evolution:

> la femme penchant son mystérieux buste de chair blanche enveloppé de dentelles ce sein qui déjà peut-être me portait dans son ténébreux tabernacle sorte de têtard gélatineux lové sur lui-même avec ses deux énormes yeux sa tête de ver à soie sa bouche sans dents son front cartilagineux d'insecte, moi? (H, 402)[28]

In this primeval dark, the child is sole fulfilment of the mother's desire.

Yet the last word of the novel, 'moi', is not the last word, because a question-mark hangs over this apparently almost successful ending to the quest for origin and identity. Can this insect really be me? The last sections of the novel do not show an untroubled homecoming into the peace of night. The narrator's drive is illuminated by the artificial lights of villages and towns which are noisy with the clamour of history (Lambert's election campaign), and his mind is laden with memories of guilt and loss. From a Freudian point of view the ending is inherently unstable, a regression to pre-Oedipal fantasies, not the necessary acceptance of castration, identification with the father and internalisation of his values. Continuing lack, uncertainty and instability of identity are signalled in the switches between first person and third, from nephew to uncle and back again. A particular source of this instability may be traced to the father's absence. To enjoy exclusive possession of the mother, the child pays a heavy price: he must dispose of the father. That price is all the greater when the father, as in the fiction of *Histoire*, is really dead. In the last chapter of the novel, where the father's death becomes explicit, grief and guilt attach themselves primarily to Hélène; but she elides with the mother; and it may be that in this palimpsest we should read the father also – a death for which the child, illogically but inescapably, feels responsible. To speak then for a moment in Lacanian terms, *Histoire* does not end in a triumphant assertion of the plenitude of the imaginary. On the contrary, the final question-mark reminds us that this is primarily a novel of desire and loss. Pugh's comment on the mother's death scene could be applied to *Histoire* as a whole: it can 'be seen to refer indirectly to the child's initiation into language, and the concomitant renunciation of the exclusive love of the mother that is the price the child pays when it enters into the symbolic relationships that are articulated in language'.[29]

From *Le Tricheur* to *Histoire* Simon's novels lend themselves to a Freudian approach. Increasingly thereafter, from *La Bataille de Pharsale* to *Leçon de choses*, there follows a period which poses problems to psychoanalytic criticism, and not simply to the Freudian form of such criticism. Celia Britton's analysis illustrates the difficulties.[30] While emphasising desire and phantasm, Britton continues the tradition of Ricardou in that she sees at the core of Simon's work a tension between realist and anti-representational elements. This opposition Britton develops in two sets of concepts. On the one hand she links realism, and above all visual representations, with the presence of a psycho-realist subject and with Lacan's Imaginary Order. That is to say that the novels from *Le Vent* to *Histoire*, and even as far as *La Bataille de Pharsale* and *Les Corps conducteurs*, encourage or at least permit readers to perceive their fictions as the phantasms of a containing consciousness. These phantasms could be described in Lacan's phrase as 'a series of alienating identifications'. Thus the novels represent the impossible desire for plenitude and fixity characteristic of the Imaginary Order: the desire for perfect models of the father, perfect union with the mother, a final explanation of origins, stable identity and continuity with the past – none of which is to be found. On the other hand Britton associates the anti-representational aspect of Simon's novels with what she calls the subject-in-language, which she describes as a subject 'existing only by implication, in so far as language requires a subject'.[31] To describe this subject more fully she describes it in terms of Lacan's Symbolic Order. For Lacan the individual's entry into the symbolic order of language creates the subject – not as commanding reflexive ego, but as lack, an incomplete signified, constantly slipping along the signifying chain. This is the 'subject-in-language' of much of Simon's fiction.

These distinctions and definitions enable Britton to say many illuminating things about the novels from *Le Vent* to *Histoire* and beyond as far as *Les Corps conducteurs* – a number of which I have already quoted. Indeed her model is particularly persuasive in showing how the balance switches from the representational to the anti-representational pole in *La Bataille de Pharsale* and *Les Corps conducteurs*. However, Britton's analysis is more tentative and her comments much sparser when it comes to the most formalist of the reputedly formalist novels, *Triptyque* and *Leçon de choses*. It is not that these novels are any less visual; on the contrary, they are more so, even although they also contain representations of the other senses – hearing, touch, less frequently smell – with which Britton's argument is not concerned. It is rather that Britton is most at ease when both poles of the tension which she sees as central to Simon's work are present; the problem with these novels is that both are apparently missing. They lack on one hand the sense of a subject, however fragmentary, striving to assert unity and identity. On the other, as Britton frankly puts it: 'the subject-in-language has been overtaken – has been so to speak leap-frogged – by a modified, depsychologized version of the subject of perception'.[32] The following extract will allow us initially to get some idea of Britton's difficulty:

> Le tireur peut encore lire facilement le titre du paragraphe imprimé en gras mais il est obligé de rapprocher le livre de la fenêtre pour déchiffrer le texte en fins caractères: Nous n'avons parlé jusqu'à présent que de l'action des eaux continentales sur les pierres et les terrains; mais l'eau des mers peut agir aussi. Le vent soufflant à la surface de la mer produit les vagues, qui viennent parfois se jeter avec violence sur les côtes. La mer entame alors les bords du continent, elle fait effondrer les roches et les terres, elle arrache les pierrres les plus dures et les roule dans ses eaux. Comme dans le cas des eaux continentales, les roches seront détruites par la mer avec une inégale rapidité, suivant leur plus ou moins grande résistance. Lorsqu'une même roche aura des parties compactes et d'autres qui seront plus molles, les premières, moins vite démolies par les vagues, formeront au milieu de la mer des colonnes ou des piliers. Ainsi à Étretat (fig. 111), à Dieppe, etc., la craie qui forme les falaises au bord de la mer a des parties très friables et d'autres plus solides. Les premières s'écroulent plus vite et c'est ainsi qu'on voit se former sur le bord de la mer de grandes arcades de rochers soutenues par les parties les plus résistantes qui en forment les piliers. Au premier plan de l'image, les lignes qui représentent la mer s'espacent et s'épaississent à la fois, laissant entre elles par endroits des vides blancs, allongés et un peu boursouflés, comme de faibles rouleaux d'écume. A l'intérieur de chaque vague, les galets qui forment le fond semblent se soulever dans l'eau transparente, comme un tapis qu'on roulerait, puis la crête de la

> vague se brise et ils réapparaissent à leur place primitive pour se soulever de nouveau. Les vaguelettes viennent mourir l'une après l'autre sur le rivage avec un bruit frais. Il semble que çà et là des reflets roses jouent sur l'eau vert pâle. Le pourvoyeur pousse le coude du tireur et dit Ho Charlot j'te cause tu m'entends merde c'est le moment de bouquiner kes'tu lis? Il lui arrache le livre des mains, oriente les pages vers la fenêtre et lit le titre en caractères gras: 145. DESTRUCTION DES CÔTES PAR LES VAGUES. Il dit merde et la destruction des cons comme nous où c'est qu'ils en parlent? Il jette rageusement le livre. Pendant que l'un des maçons maintient à hauteur de sa poitrine les extrémités des planches qu'il soulève, l'autre, accroupi au-dessous, s'efforce de tirer à lui l'un des lourds tréteaux dont les pieds sont parfois bloqués par un amas de gravats, une brique ou une pierre coincées qu'il lui faut dégager à la main. A la fin il se relève, fait signe à l'autre ouvrier qu'il peut laisser retomber les planches et sort de la pièce pour y revenir peu après, poussant devant lui une brouette de fer à la roue garnie d'un pneumatique. En travers de la brouette est couchée une pelle de terrassier au fer arrondi et ébréché. Elle sent ses moustaches sur son cou au-dessous de son oreille elle sent ses lèvres elle sent soudain sa langue humide et râpeuse sur sa peau elle frissonne, la main à plat contre sa poitrine elle le repoussse, elle dit non laissez-moi je vous défends non, elle se cambre contre la barrière et recule son visage. Dans le noir il apparaît comme un ovale imprécis, bleuâtre, dont les reliefs ne sont indiqués que par deux taches sombres marquant les cavités des yeux et une plus grande à la place de la bouche. Il jette son cigare qui tombe sur une pierre du chemin et une brève pluie d'étincelles s'éparpille dans le noir. (LC, 94–7)[33]

The tone of this passage is for the most part impersonal, its narrator unplaced, its narrative elements largely unhierarchised. Although the quotation from the textbook and subsequent description of an illustration appear to be framed by the soldiers' story since a soldier is reading the book, no such framing links that episode to the story of the masons working in a room (the same room?), or either of these fragments to the rendezvous between a man and a woman. On the face of it, then, this text offers as little hold to an untrained reader as it does to one as subtle as Britton. On the other hand, it is just the kind of text which lends itself to Ricardolian analysis. Take for example the transition from the masons to the lovers. It can be shown that continuities of language and motif subtend the referential discontinuity. The transitions are bridged by contrasting actions ('l'autre s'efforce de tirer à lui', 'elle le repousse'), by contrasting yet similar visual patterns (the spade crossing the line of the wheelbarrow, the woman's back against the horizontal of the gate), and by repeated sounds (plosive consonants and the 'ou' vowel in 'roue', 'brouette', 'couchée', 'moustaches', 'cou', 'au-dessous'). Finally, the sexual suggestiveness combined in the word 'couchée' and the phallic shape of the spade seem to motivate, or as Ricardou would put it – using the language of psychoanalysis without its content – 'overdetermine' the transition. As François Jost wittily observed: 'Dans cette optique, on tiendra l'aventure de la femme pour un simple résultat de ses mauvaises fréquentations contextuelles.'[34]

Is there any way to escape this familiar sterile scenario of representation subverted by linguistic play? Celia Britton makes a comment which helpfully points a way forward:

> whereas in the earlier texts the language itself participates in the rhythms of a desiring vision, it now [in *Triptyque* and *Leçon de choses*] appears to be refusing desire by its impersonal exactness. But this is so self-conscious – an entirely deliberate and carefully maintained exclusion – that its effect is, arguably, to reproduce desire in a different form: that is, one that serves to dramatise the mechanisms of its repression.[35]

One might gloss Britton's remark this way. In the passage quoted above, many of the traditional signs of subjectivity are, so to speak, suppressed – framing, hierarchy, personality – so that we cannot find the totalising effort of the conscious ego. Equally, the passage demonstrates what might be called 'the verbal mechanisms of repression', the gaps and leaps of associations in language, simultaneously showing and hiding the repressed. But I would suggest that desire is much less repressed than Britton supposes and that *Leçon de choses* continues the crablike advance of Simon's work. Moving forward by moving backward, it goes further in exploring the origins of the subject, and the impossibility of reaching them.

Leçon de choses begins with a dramatisation of its own origins. In the short introductory section, entitled 'Générique', Simon describes a room. In the first paragraph the writing moves metonymously with neutral precision from wall to floor and ceiling; in the second it branches into metaphors of sea and landscape. The third paragraph draws the moral and continues the process:

> La description (la composition) peut se continuer (ou être complétée)

> à peu près indéfiniment selon la minutie apportée à son exécution, l'entraînement des métaphores proposées, l'addition d'autres objets visibles dans leur entier ou fragmentés par l'usure, le temps, un choc (soit encore qu'ils n'apparaissent qu'en partie dans le cadre du tableau), sans compter les diverses hypothèses que peut susciter le spectacle. Ainsi il n'a pas été dit si (peut-être par une porte ouverte sur un corridor ou une autre pièce) une seconde ampoule plus forte n'éclaire pas la scène, ce qui expliquerait la présence d'ombres portées très opaques (presque noires) qui s'allongent sur le carrelage à partir des objets visibles (décrits) ou invisibles – et peut-être aussi celle, échassière et distendue, d'un personnage qui se tient debout dans l'encadrement de la porte. Il n'a pas non plus été fait mention des bruits ou du silence, ni des odeurs (poudre, sang, rat crevé, ou simplement cette senteur subtile, moribonde et rance de la poussière) qui régnent ou sont perceptibles dans le local, etc., etc. (LC, 10–11)[36]

Two affirmations are made here, of which the second ultimately undermines the first. First it here becomes explicit that the function of the description is to generate the novel. Everything which follows is to be traceable to a single originating act. The text claims to know and be master of its origin, to be a creation *ex nihilo* from the moment the pen is put to paper. Second, from this single act spring almost infinite possibilities of complex growth, through further description ('l'addition d'autres objets'), metaphorical shifts, reframing (the description may itself be a a picture contained within a frame), hypotheses, the introduction of senses other than sight, 'etc., etc.'. It is precisely this sense of infinite potentiality, however, which in the end puts the originating act in question. The elongated shadow on the floor, for example, implies a human being. What is this character doing there, how did he or she get there? How did the objects become worn or broken? Why does the room smell of powder, blood, or dust? The answers to all of these questions demand reference to time, to what went before; and each answer in turn will demand further explanation in an infinite regression. Thus, this paragraph which affirms responsibility for its own origin also raises questions about origins, as if no statement can be made which does not imply a previous, earlier origin.

In fact, the subsequent construction of the novel depends very largely on Simon's recognition and acceptance of an inescapable anteriority. In one sense the initial description generates everything which follows: the narrative of the soldiers trapped in the room in time of war; the narrative of the masons, building, altering, demolishing the room; the narrative of the lovers which emerges from the illustration on a calendar hanging in the room. But in a more profound sense the text explores what already exists. Building, as becomes evident from the work of the masons, is a matter of demolition and reconstruction. This metaphor for the work of the novelist makes its appearance at the very beginning of the novel in that the first words already show how language resonates with multiple meanings. 'Les langues pendantes de papier', dangling strips of wallpaper, are also the tongues, the languages of which the novel is composed, and to strip one away is to come upon another. So the text wanders in a forest of signs. At the most general level, the signs are those of language itself, the associations of sense and sound which prompt the text on its way. But the text also deals with many more specific codes and subcodes of language and culture. For example, the code of Impressionist painting – the lovers' cliff-walk is part Monet, part Boudin; or, more simply, the code of the list – the saints' days on the calendar are chopped up and carried round the text in fragments; or again, codes of language specific to particular writers: Michael Evans has pointed out how the lovers' rendezvous is a version of Emma's seduction by Rodolphe[37] (to which one might add that the woman goes to meet the man on the same pretext Emma uses with Léon, that she must tell him that she can't come); or the factual/didactic style of the textbook 'Leçons de choses' of which so many fragments are incorporated into the novel; or, casting the net more widely again, the code of a popular oral register of language: the lament of the grumbling soldier.

If we now turn back to the passage quoted earlier two points become evident. First, the surface sameness is a gloss which conceals considerable variety. The cool, confident, explicatory tone of the textbook gives way to a literary lyricism. What begins as a description of the textbook's illustration ends as an evocation, glinting with metaphor, of the movement and sound of waves tumbling and rippling on a beach. This in turn is rudely contrasted with the soldier's speech which interrupts it, a contrast reinforced by the momentary departure from conventional punctuation and spelling: 'j'te cause tu m'entends merde c'est le moment de bouquiner kes'tu lis?' There follow two contrasting narrative codes in the third person: behaviouristic in the account of the masons, with shifting internal focus for the lovers: first from her point of view (the repetition of 'elle sent' and the sensations of touch – moustache, lips and tongue – give psychological colouring to the scene); then from his, as her face becomes indistinct in the shadows.

Second, it becomes clear that one origin of the text, indeed its principal origin, is Simon's own previous work. It is not hard to discern here many of the same themes which we have been following throughout this chapter. Destruction and violent death are once more linked to creation and procreation: the cliffs erode, the soldiers die, the masons build, man and woman come together. Once again the filtered, ordered certainties of the written word are rejected and denounced: 'merde et la destruction des cons comme nous où c'est qu'ils en parlent?' Delight in the sights and sounds of life, in the recurrent pattern of the waves, is tempered by the sense of potential loss reflected in the woman's reluctance to embark on the cycle of procreation: for the individual the cycle ends in death.

The single missing element might appear to be the explicit fantasies of origin associated with the family. In fact, these are also present. They undergo in *Leçon de choses* the same process of splitting and grafting, of fragmentation, destruction and reconstruction which affects all other themes. For example, the uncentred self which was almost Georges in *La Route des Flandres*, which split between narrator and uncle in *Histoire*, and which fragmented into multiple viewpoints in *La Bataille de Pharsale*, is now scattered and reassembled in new configurations: a version of the events of Georges's débâcle is recounted by the older mason in the language of the jockey Iglésia, Georges's situation and emotions echoed in the plight of the young soldiers, abandoned, famished, facing destruction at the hands of an enemy vastly superior in numbers and equipment. More specifically, the drama of origins resurfaces in the encounter between man and woman. Some four pages after the passage quoted, and chronologically later in the narrative, the young woman unpins a medallion which holds her collar in place to allow the man to slip his hand into her corsage and reveal one of her breasts: 'il serre et deserre sa main ou frôle de la paume le mamelon élastique. [. . .] Sans cesser de la caresser, il regarde la coulée de la chair laiteuse' (LC, 101).[38] This series of actions reworks and transforms a fragment in *La Bataille de Pharsale*. In the section entitled 'César', the narrator recalls a memory from his first visit to Lourdes as a young boy. Reaching for her purse to tip the hotel porter, his grandmother undoes her corsage, similarly unpinning a medallion to reveal the whiteness of her breast:

> (cette poitrine que j'entrevoyais pour la première fois, découvrant que sous les mystérieuses soies noires se trouvaient une peau, des chairs dont la mollesse, l'extrême blancheur, accentuaient encore l'irréalité, le caractère sacré de choses destinées à rester cachées et sur lesquelles – de même que l'hostie au moment de l'élévation – on ne doit pas porter les yeux). (BP, 125)[39]

In this passage the biblical language and comparison with the host heighten the horrified fascination of that which is forbidden. The link between the two passages is all the clearer because the incident in *La Bataille de Pharsale* prompts the narrator to reflect on photographs of his grandmother as a young woman, 'les robes, les coiffures démodées' (BP, 125); her youth corresponds to the time which Simon evokes in the dress and manners of the lovers' narrative in *Leçon de choses*. Other components of the incident from *La Bataille de Pharsale* are differently reworked, rearranged. In *La Bataille de Pharsale*, transgression of a sacred law provokes guilt and shame in the narrator; in *Leçon de choses* these feelings are transferred to the woman, while the man, unusually for Simon, is not victim but dominant seducer. There is a moment in *Histoire* when Simon's language fuses image to image with such charged intensity that it ceases to have the representational effect of prose. Celia Britton, drawing on a number of textual parallels, has suggested that this confused explosion of language coincides with the text's approach to that which cannot be acknowledged, the unconscious desire for the mother.[40] One might argue conversely that in *Leçon de choses* rearrangement of the narrative elements of *La Bataille de Pharsale* distances this same theme and permits this detailed representation in untroubled prose of a sexual act which runs its full course. This intertextual reading reveals the continuing presence of the incestual element, passing from mother/model in *Histoire*, through grandmother in *La Bataille de Pharsale* to grandmother as young woman in *Leçon de choses*. Thus *Leçon de choses*, it may be said, realises in distanced form what in earlier novels was repressed: the fantasy of incest. This incestual colouring gives allegorical expression to the ambition declared in 'Générique': it is a fantasy of origin but equally a fantasy of self-origin.

It is possible to see *Leçon de choses* as figuring the drama of human accession to the Symbolic Order: to use language is to jump on a train which is already in motion and which prevents the self from ever establishing fixed contours or a stable origin. *Leçon de choses* lacks however the rending and grinding which Lacan associates with this process. It would be more appropriate to see earlier novels, from *Le Vent* to *Le Palace*, in this light, novels in which the subject wrestles with language, adrift in the current of the Symbolic Order but fighting against it to preserve or restore the plenitude of the Imaginary. *Leçon de choses* continues the tendency visible in *Histoire* to co-operate with the Symbolic Order. The subject here is like a fish in water, revelling in it, appearing now here now there, at ease in a element which defies the human observer's attempts to measure shape and distance. *Leçon de choses* may better be understood in the light of Lacan's reinterpretation of Freud's gnomic dictum: 'Wo es war, soll Ich werden.' Conventional wisdom, in particular the standard French translation of this sentence, holds Freud to mean that the territory of the Id should be progressively invaded and captured by the Ego: 'Le moi doit déloger le ça.' Lacan understands the sentence quite differently, almost, among other senses, to mean the reverse: we require to recognise that in the Id, which can be known only in language, in the network of signifiers, the subject is already present: 'Ici, dans le champ du rêve, tu es chez toi.'[41] This reversal would be one way to describe Simon's development in the period I have been most concerned with here, from the 1950s to the 1970s. The reconstructive mode presides over novels of conflict in which the Ego seeks in vain to impose its will on the Id. Progressively, as language is allowed freer rein, the Ego begins to find its home in the Id, and in dreams of origin.

7

L'Acacia: myths of history, family and self

L'Acacia, published in 1989, is a novel of echoes. In twelve chapters Simon interweaves the stories of two families and two men. The two families differ in history, in values and in social class, but the son of the near-illiterate mountain farmer becomes an infantry officer and marries the descendant of a general. The officer goes to war in 1914, as does his son in 1939. The father is killed at the Front, the son returns. Focusing alternately on the two men, Simon multiplies parallels and contrasts, shades of difference and similarity, in situation, tone and style. Not chronological time but these shifting patterns order and link successive chapters. Father and son, mother and aunts, arrivals and departures, journeys by train and ship, the hardships of war, grief, fear, exultant joy: each character, incident and emotion plays against and recalls others. When Simon touches a string the harmonics of the note sound in sympathy. This much is available to every reader. To those familiar with others of Simon's novels, however, *L'Acacia* resonates with even greater force. Here Simon reworks his entire *oeuvre*. Not merely does he for the first time bring together the two main families of his work and compose a fuller portrait of the officer father sketched in *Histoire*; in addition he reintroduces and gives new colouring to characters, episodes and motifs from all his previous novels. *L'Acacia* rustles with the echoes of other leaves, other summers. Set your ear to any part of it and you hear the murmur of distant voices. In an article admirably bringing out the richness of intertextual resonance, Ralph Sarkonak remarked that 'déchiffrer *L'Acacia* est une expérience de grande jouissance.'[1] This hits the mark. *L'Acacia* is a work of integration and celebration. The question to be addressed is that of meaning. In what sense does *L'Acacia* integrate all that went before? What does it celebrate?

I intend to approach the novel through two extracts. The first is from an interview with Simon, one of a series of eyewitness accounts of the débâcle of 1940 published in *Le Figaro* in 1990 to commemorate the fiftieth anniversary of the Fall of France.

> **– C'est donc bien là le colonel que vous décriviez dans 'La Route des Flandres'. Il existait donc?**
>
> – Oui, et je vous renvoie à mon récit: la pagaille, ceux qui étaient devant, refluant, et ceux qui étaient derrière, voulant passer, l'ordre stupide donné par notre lieutenant: 'Combat à pied!' (c'est une manoeuvre impossible à exécuter sous le feu de l'ennemi qui pourrait faire alors un massacre: comme quoi vous pouvez voir que la totale incurie du commandement régnait à tous les échelons de la hiérarchie, des généraux aux colonels en passant par les lieutenants – il faut toutefois rendre justice à notre général: il a eu l'élégance de se brûler honorablement la cervelle . . .), puis, aussitôt après, le contre-ordre: 'A cheval et au galop!', et alors ma sangle trop longue pour la petite jument que l'on m'avait donnée en remplacement de celle que j'avais crevée pour repasser la Meuse, ma selle qui tourne au moment où je mets le pied à l'étrier, ma course en tenant la jument par la bride, le choc (un cheval, le souffle d'un obus?) qui me fait perdre connaissance, et quand je reviens à moi, je me trouve à quatre pattes au milieu d'un chemin, entouré de chevaux et d'hommes morts ou blessés. Ma course de nouveau vers la haie la plus proche, comment j'ai réussi à passer entre les blindés qui patrouillaient, ma marche en forêt en direction de l'ouest, de la ligne fortifiée que je savais être là et où j'imaginais qu'on allait m'accueillir avec des félicitations, et quand j'y suis enfin parvenu, stupeur: le silence, rien qu'un lointain bruit de canon ou d'explosions, l'éblouissante nature printanière, les chants des petits oiseaux et . . . *PERSONNE*! De cela, je peux témoigner: le matin du 17 mai 1940, il n'y avait *PERSONNE* dans les ouvrages défensifs au nord de Solre-le-Château! La ligne fortifiée avait été purement et simplement abandonnée sans combat, intacte, sans la moindre trace d'un obus ou d'un bombardement quelconque.[2]

The second extract is from *L'Acacia*.

> De même qu'il ne pourrait pas non plus dire combien de temps il est resté sans connaissance sur ce que l'on pourrait pas appeler exactement un champ de bataille (le carrefour de deux chemins vicinaux au milieu de blés en herbe et de prairies en fleurs): tout ce dont il se souvient (ou plutôt ne se souvient pas – ce ne sera que plus tard, quand il aura le temps: pour le moment il est uniquement occupé

à surveiller avec précaution le paysage autour de lui, estimer la distance qui le sépare de la prochaine haie, tandis qu'il fait passer par-dessus sa tête la bretelle de son mousqueton, ouvre la culasse, la fait basculer et la retire) ce sont des ombres encore pâles et transparentes de chevaux sur le sol, un peu en avant sur la droite, tellement distendues par les premiers rayons du soleil qu'elles semblent bouger sans avancer, comme montées sur des échasses, soulevent leurs jambes étirées de sauterelles et les reposant pour ainsi dire au même endroit comme un animal fantastique qui mimerait sur place les mouvements de la marche, la longue colonne des cavaliers battant en retraite somnolant encore au sortir de la nuit, les dos voûtés, les bustes oscillant d'avant en arrière sur les selles, la tête de la colonne tournant sur la droite au carrefour, puis soudain les cris, les rafales des mitrailleuses, la tête de la colonne refluant, d'autres mitrailleuses alors sur l'arrière, la queue de la colonne prenant le galop, les cavaliers se mêlant, se heurtant, la confusion, le tumulte, le désordre, les cris encore, les détonations, les ordres contraires, puis lui-même devenu désordre, jurons, s'apprêtant à remonter sur la jument dont il vient de sauter, le pied à l'étrier, la selle tournant, et maintenant arc-bouté, tirant et poussant de toutes ses forces pour la remettre en place, luttant contre le poids du sabre et des sacoches, les rênes passées au creux de son coude gauche, bousculé, se déchirant la paume à l'ardillon de la boucle, assourdi par les explosions, les cris, les galopades, ou plutôt percevant (ouïe, vue) comme des fragments qui se succèdent, se remplacent, se démasquent, s'entrechoquent, tournoyants: flancs de chevaux, bottes, sabots, croupes, chutes, fragments de cris, de bruits, l'air, l'espace, comme fragmentés, hachés eux-mêmes en minuscules parcelles, déchiquetés, par le crépitement des mitrailleuses – puis renonçant, se mettant à courir, jurant toujours, parmi les chevaux fous, les cris, le tapage, la jument qu'il tient par la bride au petit galop, la selle sous le ventre, puis soudain plus rien (ne sentant même pas le choc, pas de douleur, même pas la conscience de trébucher, de tomber, rien): le noir, plus aucun bruit (ou peut-être un assourdissant tintamarre se neutralisant lui-même?), sourd, aveugle, rien, jusqu'à ce que lentement, émergeant peu à peu comme des bulles à la surface d'une eau trouble, apparaissent de vagues taches indécises qui se brouillent, s'effacent, puis réapparaissent de nouveau, puis se précisent: des triangles, des polygones, des cailloux, de menus brins d'herbe, l'empierrement du chemin où il se tient maintenant à quatre pattes. (L'A, 88–90)[3]

In the first place, the two passages are alike in subject matter. They describe, as it were, the same events in different ways. Simon as historical witness does not differ factually from Simon as novelist. This emphasis on fact is typical of the novel as a whole. Simon carries further here the autobiographical project I defined, tentatively, in Chapter 5. Or, as Simon prefers to put it, *L'Acacia* is a novel 'à base de vécu'.[4] In general and in detail Simon took pains to ensure that it should, as far as possible, be factually accurate. In 1919 Claude Simon was taken by his mother and his paternal aunts through the devastated countryside of eastern France to search for his father's grave.[5] This episode forms the basis for Chapter 1 of *L'Acacia*, entitled '1919'. Equally based on Simon's life are the experiences attributed to the corporal in *L'Acacia*, his pre-war journeys, his mobilisation, wartime marriage, experience on the Flanders roads and in prison camp, escape from a camp in the Landes, and return to a southern French town in late 1940. The corporal's family-tree closely resembles that of Simon; here too Simon has shown respect for verifiable truth. In portraying the officer, fictional counterpart of his own father, Simon draws on family tradition, on photographs, and surviving documents. In Chapter 7 of *L'Acacia*, '1982–1914', the ex-corporal consults two elderly relatives about his father's past; in 1982, two non-fictional versions of these cousins of Claude Simon were living in Perpignan. In Chapter 3, '27 août 1914', Simon describes the ceremony at which a decoration was awarded to the officer's regiment. This description is based on the text and photographs contained in a brief history of the Great War service of the Twenty-fourth Regiment of Colonial Infantry.

Simon's resolve to 'give up fiction' in *L'Acacia* gives direction to his rewriting of earlier novels.[6] Sometimes he provides context and explanation: the story of 'l'homme-fusil' from *Le Palace* was told him by an Italian anarchist on the train to Barcelona; the uncle of *La Corde raide* is now firmly attached to his mother's side of the family, a version of the younger of her two cousins. With *La Route des Flandres* Simon deals more radically. His corrected version of that novel filters out the fiction, with zest and ironic humour. Much of the earlier novel remains: the corporal escapes from the massacre of the squadron, flees on foot, alone, through fields and woods, is reunited with the two officers and the batman, sees his commanding officer's sabre glint in the sunlight as the sniper guns him down. But the imaginings which in *La Route des Flandres* Simon wove round this episode are swept away. The commanding officer is no longer a distant relative, no longer a captain (the two most overt signs in *La*

Route des Flandres of his status as father substitute), but instead the regimental colonel.[7] Although the officer's batman was a jockey, like Iglésia in *La Route des Flandres*, so too were half the soldiers in the regiment. And this particular jockey, member of the apocalyptic four on the Flanders road, the corporal had never seen before. What is more, the soldier–jockey whom the corporal knew best, whose Italian-sounding name and distinctive mode of expression might make one think of Iglésia, had no connection with that officer, was not taken captive to Germany, indeed was not even present at the débâcle: the day after war was declared he broke his ankle and was sent home. As for the sexual fantasies associated with Corinne, Simon would now have us believe that, while the German guards had lively sexual interests (the corporal's obscene drawings sold like hot cakes), the regime of the prison camp left men with only two desires, food and escape. Only when the latter was achieved did the sexual urge return. The vain attempt to tell the story to someone who had not experienced and could not understand it was made not to the elegant widow of a fallen officer, but to a no less desirable prostitute in a third-class brothel in the corporal's home town.

Paradoxically, this ironing out of fiction makes the stories of *L'Acacia* no less like fiction in the sense that events are just as dramatic, extraordinary, constructing almost of themselves their own ironies. Only a novelist enamoured of fiction, one might have thought, would have risked reuniting the corporal and his Jewish friend (the new version of Blum) in the same horse van on the same prison train when the former had spent ten days traversing and re-traversing Belgium in every direction while the latter, returning from leave in Paris, was simply greeted at the platform by a German reception committee and invited to change trains. Who would have dared invent the coincidence that the corporal should leave for the front from the same town and the same station as his father, twenty-five years to the day after his father died in the same north-eastern provinces to which he was now heading? Simon has discovered the trick which eluded him in his early novels: the only successful way to portray chance in fiction is to persuade your readers that what they are reading is not fiction, or is at least fiction based on fact. Most extraordinary of all, when one considers his opportunities to die, the corporal survived the war. What *L'Acacia* celebrates most obviously is that, against all odds, the corporal was spared, escaped and resumed his life.

The novel is ordered so as to bring this out: the first chapter dwells on the captain's death, the last chapter on the corporal's survival. This exemplifies a truth everywhere apparent in *L'Acacia*: Simon manipulates factual accuracy. The precise dates and days of the chapter headings, for example, are doubly functional: they place events in the context of historical time, while simultaneously emphasising how much time is excluded. Some omissions may be partly due to lack of knowledge. It appears Simon did not know that his newly-married parents spent two years in Perpignan before leaving for Madagascar.[8] But how much more symbolically apt that the mother's change of status, her sensual awakening, should be accompanied by the exotic discoveries of travel. Nor did Simon know, or choose to acknowledge, that Captain Antoine Louis Eugène Simon's death certificate, held at the town hall in Arbois, states explicitly that his father died at 'la forêt de Jaulnay'. It is therefore unlikely, but how much more effective in conveying the difficulty of the search for the grave, that the widow should not know whether she was looking for Jaulnay, Gaulnay or Goulnoy (L'A, 18).

Other manipulations are conscious and planned. As in *L'Herbe*, Simon simplifies his father's side of the family by conflating three sisters into two, more precisely by blending two of the sisters into the older sister of *L'Acacia*, that outspoken, mannish, fiercely anti-clerical tiller of the soil. The omissions from the fictional chronicle of Simon's own life are innumerable. The subject dictates that there should be little of the Spanish material. More striking is the jump from young boy (Chapters 1 and 11) to young man; with the partial exceptions of *Le Tricheur* and *Histoire*, Simon is not a writer of childhood. Even *L'Acacia*'s elaborate account of the débâcle excludes key scenes from earlier works, for example the plunge down the railway embankment and the cavalryman's subsequent escape on foot along the track (*La Corde raide*, *La Route des Flandres*, *La Bataille de Pharsale* and *Leçon de choses*). A simple reason suggests itself: this incident took place on 12 May, not one of the days named in the chapter headings of *L'Acacia*.[9] More pertinently, its epitomising function as the moment when the corporal ought to have died, lost consciousness but then regained it, is played in *L'Acacia* by the incident from 17 May quoted in the passage at the head of this chapter.

In short, Simon selects and organises; *L'Acacia* is constructed; what Simon constructs is a myth of history, of family and of self.

Before *L'Acacia*, Simon uses the word history with a capital letter in all his works from *La Corde raide* to *Histoire*, and again, more than twenty times, in *Les Géorgiques*. Many of his critics, from Ludovic Janvier onwards, have been prompted to raise questions about what history means for Simon.[10] What are the relationships between the meaning he gives to the word and the themes of his novels, between these themes and twentieth-century history, between how his novels are written and how history is written? One possible key to Simon's conception of history is to think of it primarily, though not exclusively, as absence. Georges strives desperately to fill the gaps in his knowledge of the past, to find fullness of explanation and final truth; Blum pours satirical scorn on those who believe they possess such knowledge. Were history to exist, it would be a transcendent certainty giving meaning to time and ground to identity. For Simon no such certainty exists. Yet he wishes it did. Hence history is a void which cries out to be filled. Simon pours meanings into the word, but all such meanings are fragmentary, confused, even self-contradictory, shadows of what the substance of history could have been. That lost substance appears in the forms of parody, caricature and satire.

For example, Christian providence appears in Simon as a distorted mirror-image of itself. The priest in *Le Tricheur* is caricatured for his tranquil confidence in the goodness of God's plan. Montès in *Le Vent* is a Christlike figure, rejected by society; his suffering brings no redemption. In *Histoire* Simon glosses sceptically the mother's confident expectation that in death she will be reunited with her husband. In place of providence, the expression of the loving care of a personal God, Simon frequently personifies history as either an indifferent or a malevolent force, bringing not salvation but catastrophe. In *Les Géorgiques*, O. and the soldiers of 1940 are caught up in a meaningless ritual presided over by such a destiny: 'peut-être, après avoir frappé un premier coup, epouvantée elle-même, l'Histoire s'accordait-elle un répit' (LesG, 131); l'Histoire elle-même se chargeant du reste [. . .] agissant (l'Histoire) avec sa terrifiante démesure, son incrédible et pesant humour' (LesG, 340).[11]

A second version of history, glimpsed and rejected by Simon, is a form of rationalist humanism. Celia Britton has analysed, in the ideas defended by Pierre in *La Route des Flandres*, the traces of the philosophers of the Enlightenment and of Rousseau.[12] To the extent, however, that one can reconstruct a hypothetical fullness of meaning to this version of history, it might seem rather to encompass both the eighteenth-century and nineteenth-century positivism. Positivist historiography, in the flush of its nineteenth-century confidence, believed that the past was accessible to reason and hence to the historian. By the painstaking work of collating and verifying his sources, he could reconstruct events and establish laws of cause and effect. Underlying this confidence were beliefs and assumptions inherited from the past: from Newton and Locke, the conviction that the world is subject to constant laws and that man passively receives sense impressions from the external world; from the writers of the Enlightenment that human nature too is constant; and from Christianity, via the Enlightenment, a secularised version of belief in the linearity of history and optimism about the future: in short, a belief in progress, moral and material.[13]

Simon rejects this model of history comprehensively. Blum remarks that 'Gorillus sapiens', distancing himself from Christian belief, has set all his hopes on happiness, to be obtained by material progress ('la production en grande série de frigidaires, d'automobiles et de postes radio', RF, 188[14]); but the elevated position from which the gorilla views history is the top of a rubbish dump. What the past leaves behind, physically and metaphorically in Simon, is decaying fragments. The hope of material progress is derided in the burnt-out wreckage which strews the Flanders Road, in the gimcrack standardised products of modern urban civilisation and in rusting farmyard machinery: the reaper–binder of *La Bataille de Pharsale*, designed to function with utilitarian purposefulness, has definitively seized up. As for the hope of moral progress, its manifestations are satirised from the Enlightenment to the Third Republic. The Thomas family of *L'Herbe* are spiritually descended from the ancestor generals of *La Route de Flandres* and *Les Géorgiques* since positivism inherits the ancestors' faith in virtue, reason and human perfectibility, and adds to it the expectation of social progress. As we saw in Chapter 1, the ascension of the Thomas family – its escape from the drudgery of the land, its conquest of knowledge – is arrested in Georges's disillusionment and the catastrophe of war.

Simon's attitude to a third version of history is fiercely ambivalent, as if, positively and negatively, it engaged him most. To the first, materialist premiss of Marxist history, Simon is firmly attached, 'the premiss, namely, that men must be in a position to live in order to "make history". [. . .] The first historical act is thus the production of the means to satisfy these needs, the production of material life itself.'[15] Simon's rejection of idealisms, whether Christian or humanist, his insistence on the paramount importance of fulfilling material needs, emerges most strongly in *La Route des Flandres*: in the horse's discovery of metaphysical absence and in the howl of rage which Georges directs to his father from prison. Simon's materialism also takes milder forms: Marie's exemplary refusal to be distracted into metaphysical speculation (L'H, 61–2), Cézanne's willingness to take and show the world as it is (CR, 117). This attitude casts light on other personifications of history in Simon. In castigating the idealism of the young Hegelians, Marx criticised them for using the word 'history' as if it were 'a person apart, a metaphysical subject of which the real human individuals are but the bearers: "History will not be joked at . . . history has exerted its greatest efforts to . . . history has been engaged." '[16] Simon's use of the word often echoes this practice, but as aggressive parody. He personifies history as a spirit, as a theatre director acting with heavy-handed humour, capable of splitting itself in two, shocked by its own excesses. Simon's grotesque personifications are always hypothetical, very often introduced by 'comme si'. History is an empty figure. To think that it has a will of its own is an idealist illusion.

Yet of all that is built on Marx's materialist premiss, of Marxism as a system which explains history, Simon's novels are profoundly sceptical. Through the disillusionment of exemplary characters and the form of his novels, Simon puts in question the belief that revolution may effect meaningful change. As long ago as 1955, in the short story *Babel*, as recently as 1987 in *L'Invitation* he throws sidelong disbelieving glances at the Soviet Union. In *La Corde raide*, *Le Sacre du printemps*, and particularly in *Le Palace* he elaborates themes of *naïvete* and disillusionment focused on Spain. The cynical American of *Le Palace* mocks the determinism in Marx's view of history: 'cette sacrée chère bonne vieille femme à barbe qui a tout prévu' (LP, 141).[17] Marx had not foreseen that the revolution would abort. A good motto for the unchanged reality of Spain, the American suggests, would be 'ORINA – ESPUTOS – SANGRE'.[18] His evidence lies in the assassination of the revolutionary leader. Was it the work of the Franquist Fifth Column, or an act of internecine warfare between the Revolutionaries, communists against anarchists? The fiction of *Le Palace* leaves unanswered questions. Who killed the *commandante*? What happened to the American? The reader is left with a strong suspicion that the American, like the *commandante*, was disposed of for political reasons. The symmetrically arranged chapters, the lists of disparate objects borne out and into the hotel, the trams circling round town on their set routes: all bear out the sense of change within changelessness which is conveyed by the novel's epigraph from Larousse: 'Révolution: Mouvement d'un mobile qui, parcourant une courbe fermée, repasse successivement par les mêmes points.' This geometrical definition of revolution seems to comment sceptically on revolution as a political phenomenon.

Les Géorgiques amplifies this critique of Marxist practice and belief. The effect of including two revolutions, Spanish and French, in one novel is to present history as an unchanging series of variations on the same themes. O.'s *naïveté* and blindness have to do with attempting to use Marxist literature – 'quelque chose comme de la nitroglycérine sous forme de papier imprimé' (LesG, 318) – as practical handbooks. They do not prepare him to understand 'un monde où la violence, la prédation et le meurtre sont installés depuis toujours' (LesG, 318),[19] nor in particular the violence of one section of the working class against another. The French Revolution and its aftermath confirm the cyclical movement of history. For Marx, the French Revolution does away with feudalism and signals the triumph of the bourgeoisie and of capitalism. Although capitalism exploits the working class, the Revolution is essentially a positive step, ushering in that stage in history which precedes the communist revolution. By contrast, Simon emphasises failure and circularity. L.S.M.'s memorandum of 1793 urges the Convention to vote for the execution of Louis XVI. It pleads in heroic terms for change, for the Republic and for freedom from 'les idées superstitieuses' (LesG, 201). This contrasts with the petty concern for property as, twenty years later, L.S.M.'s royalist wife and his son fight over his inheritance, in solicitor's letters complete with pious references to God's protection (LesG, 383). But even before this, in the Terror, the Revolution had lost its way, its leaders destroying one another, just as in Spain. In 'The Eighteenth Brumaire of Louis Bonaparte' Marx pointed out the difficulty of escaping the past: 'the Revolution of 1879 to 1814 draped itself alternately in the Roman republic and the Roman empire',[20] while the actors of 1848–51 aped the gestures of 1789–95. Yet Marx distinguished between these two sets of events.

The Revolution accomplished real change in its Roman garb and the Terror was a tragic aftermath. In 1848 to 1851, however, the facts and personages merely parodied those of the great revolutionary period; tragedy repeated itself as farce. Simon similarly draws attention to the Roman iconography of the Revolution, for example, in the opening sketch of the two soldiers, in the general's bust, in women's dresses and names. But in the context of *Les Géorgiques*, the Revolution's choice of clothing merely adds force to the idea of recurrence: history as a whirligig of republics and empires, Roman and French. Simon even adapts Marx's theatrical metaphor and applies it to the very epoch which Marx called tragic: the Terror itself accelerates repetition to the point of farce:

> L'Histoire [. . .] se mettant à fonctionner à vide, emballé, tournant à la parodie, au bouffon [. . .] l'invisible metteur en scène pressé d'en finir, accablé par les redites d'une pièce cent fois jouée, laissant à peine aux acteurs le temps de lancer leur réplique, faisant déjà signe au suivant, tyrans, despotes pour un mois, une semaine, un jour, morts le soir d'après. (LesG, 385)[21]

But Simon is not merely anti-Marxist, he is profoundly un-Marxist. His novels accord special prestige to phenomena which he describes as existing as if outside or prior to history: gypsies in *Le Vent* and *Les Géorgiques* ('la perpétuation, la délégation vivante de l'humanité originelle, inchangée', LesG, 208), Spain itself ('une sorte de fruit desséché et ridé, oublié par l'histoire', LesG, 320).[22] War puts its participants in touch with the elemental. It is both an age-old activity uniting present generations with past – the cavalrymen of 1940 with 'les vieux lansquenets, reîtres et cuirassiers de jadis' (RF, 30)[23] – and it brings those who participate close to nature, to earth and to the seasons. For naked unaccommodated man, concerned only with the essentials of survival, the earth-shaking events of public history pale before the enchantment of direct contact with the physical world: the drumming of horses' hoofs, the colours of a sunrise, the call of a cuckoo. So immediate, so real are such experiences that they seem almost to transcend the material. Conversely, the simplest accumulation of everyday events, as in Marie's account-book in *L'Herbe*, can be shot through with a sense of awe when set against the epic turning of the seasons. In place of the downward spiral of history, one might almost speak here of a virtuous cycle, since universal decay is complemented by irrepressible growth:

> une sorte de vie secrète, exigeante, impérieuse, comme ce qui forçait les insectes à tournoyer sur place, suspendus, emmêlant sans fin, sans but, les invisibles trajectoires de leurs vols, leur nuage suspendu, immobile, obstiné, chacune de ses particules en perpétuel mouvement, claires, dorées au-devant du fond de verdure, puis (le même nuage, ou un autre) se détachant en sombre sur le ciel pâlissant. (L'H, 236–7)[24]

Here Simon comes close to filling the void of history. This version of history is grounded in the analogy with nature; it is static in total; the parts whirl in aimless movement; but the whole is endowed with a pulsing mysterious energy, constantly self-renewing.

For Marxists, and for most non-Marxist historians, such a view of history is quite ahistorical. It does not concern itself with change. It treats nature as an unchanging universal, whereas for Marxists and others nature is a product of history and has its own history.[25] Lastly, it sees the universal in the particular, and the particular in the universal, ignoring the specific and the typical. It is in something approaching this sense that the word is used, provocatively, as the title of *Histoire*. History is the ordinary and the personal, a day spent on the minutiae of clearing up an estate, in business engagements and casual encounters, but also a day on which loss, grief, the search for identity are given universal force through the authority of images drawn from nature and classical myth. *L'Herbe* is written from the same perspective. Marie's humdrum life can be 'la matière même de l'Histoire' (L'H, 36) because her everyday resilience embodies the life force and provokes Simon's vision of the epic in the ordinary.

Yet even this is not the complete story of Simon's history. For, although in both *L'Herbe* and *Histoire* the traditional matter of history – public events, wars, revolutions – are no more than distant echoes, both also give accounts of social history in the stories of two contrasting families under the Third Republic. These two families bring us to *L'Acacia*. In this novel are to be found traces of all Simon's versions of history, shadow and substance. But it is particularly in the interlocking of private and public histories, family and society, that *L'Acacia* strikes a new note in Simon's fiction.

To study history in *L'Acacia*, we can begin by further comparing the two extracts quoted at the beginning of this chapter. How do they differ in the way they are told? The *Figaro* passage is recognisably Simon: in the first line the reader is referred not to reality but to a previous narrative. Thereafter the account proceeds not as a sequence of events linked by cause and effect but as an accumulation of discrete images, introduced by nouns and present participles, and interrupted by parenthetical comments and elaborations. When a possible cause is identified, it comes in hypothetical form: 'le choc (un cheval, le souffle d'un obus?) qui me fait perdre connaissance'. Nevertheless, this is not Simon in full flight. The characteristics are attenuated. The account is linear, chronological. The tenses are those of classic French story-telling: beginning with the past, switching to the present at the height of the action ('ma selle qui tourne au moment où je mets le pied à l'étrier'), eventually returning to the past tense and thus re-establishing a conventional distance between the past of the action and the present of the enunciation. Simon's use of the first person is equally conventional. It affirms continuity and identity through time. The person who replies to the interviewer ('Je vous renvoie à mon récit') is apparently the same as the person who lived through these events: 'Je reviens à moi.' *La Route des Flandres* is the story of the impossibility of writing that last phrase.

By comparison, enunciation in the passage from *L'Acacia* is complex. The narrative, typically, doubles back on itself, beginning and ending with the moment of unconsciousness. Or rather, perhaps it is linear: not the chronological succession of events, but the textual elaboration of memories. The opening is almost Balzacian: a voice speaks with authority about the limitations of the character's knowledge: 'il ne pouvait pas non plus dire combien de temps il est resté sans connaissance sur ce que l'on ne pouvait pas appeler exactement un champ de bataille'. But when Balzac tells you that a character does not know something, you are in no doubt that the narrative voice does know and could tell you if it chose to do so. Here, by contrast, the narrator's authority is limited to talking about limitations; and the air of authority evaporates as conventional time-distinctions erode: 'tout ce dont il se souvient (ou plutôt ne se souvient pas – ce ne sera que plus tard, quand il aura le temps: pour le moment il est uniquement occupé à'). These future and present tenses introduce new layers of perspective: what was perceived then, what was later reconstructed in memory. But more than that, these layers are not distinct and they are subsumed in the present of the enunciation: avoiding the traps of set expressions ('pas exactement un champ de bataille'), the narrative corrects itself, advances by successive approximations ('pas exactement', 'ou plutôt'). The moments which precede and follow the corporal's loss of consciousness cannot be measured in clock-time, but are infinitely expandable in the present of writing.

Third-person narrative is Simon's most flexible instrument in *L'Acacia*. Here and often elsewhere in the text it is employed in prose almost indistinguishable from that of late 1950s and early 1960s to convey the texture of memory and perception. We are not however invited to read the novel as the contents of a single consciousness. Although taking the perspective of the corporal more frequently than any other, Simon also adopts the point of view of the mother (Chapter 3) and, more briefly or passingly, of a concierge (L'A, 219), the elder sister (L'A, 310), the mother's mother (L'A, 279) and the cavalrymen's collective perception of the colonel (L'A, 47). At the other extreme Simon uses third-person narrative in a form inherited from the novels of the 1970s, without trace of enunciation, as in the description of the dead officer propped against a tree:

> C'était un homme d'assez grande taille, robuste, aux traits réguliers, à la moustache relevée en crocs, à la barbe carrée et dont les yeux pâles, couleur de faïence, grands ouverts dans le paisible visage ensanglanté fixaient au-dessus d'eux les feuillages déchiquetés par les balles dans lesquels jouait le soleil de l'après-midi d'été. Le sang pâteux faisait sur la tunique une tache d'un rouge vif dont les bords commençaient à sécher, déjà brunis, disparaissant presque entièrement sous l'essaim de mouches aux corselets rayés, aux ailes grises pointillées de noir, se bousculant et se montant les unes sur les autres, comme celles qui s'abattent sur les excréments dans les sous-bois. (L'A, 61)[26]

The objectivity of this description, tender only in its precision, seems to convey certainty of knowledge – until that certainty is questioned in Chapter 11. The witnesses on whose evidence it is based may have been influenced by psychological factors (the desire to comfort the widow), or by clichéd images ('adossé à cet arbre comme un chevalier médiéval ou un colonel d'Empire', L'A, 327), or stereotyped expressions ('la balle "reçue en plein front" ', L'A, 327).[27] Simon also uses the third person in synthesised character descriptions: 'On aurait dit qu'elle n'avait pas de désirs, pas de regrets, pas de projets. Elle n'était ni triste, ni mélancolique, ni rêveuse. Plutôt gaie, racontèrent ceux qui la connurent à cette époque, gourmande (et donc sans doute sensuelle)' (L'A, 114).[28] The certainty of affirmation is attenuated here by the use of negatives, by the acknowledgement of sources ('racontèrent ceux'), and by signs of approximation ('On aurait dit', 'plutôt gaie'). The narrative proceeds from the relatively known to the relatively unknown by way of reasoned, tentative hypothesis: 'gourmande (et donc sans doute sensuelle)'. Such tentativeness in the construction of character is typical of *L'Acacia*: 'Elle était pieuse. Du moins elle accompagnait chaque dimanche sa mère et sa soeur à la cathédrale' (L'A, 117); 'Elle jouait au tennis (du moins une photographie la représentait-elle sur un court)' (L'A, 119).[29] *L'Acacia* then sets fullness of knowledge out of reach. The intertextual fragments are reworked into a new whole; but their integration is not complete or definitive, but by implication provisional, open to further reworking. Simon's critique of positivism remains intact.

Nevertheless there is a shift of emphasis in *L'Acacia*, even in comparison with *Les Géorgiques*. The previous novel began with an act of homage to its intertextual origins: the two soldiers are representations of a representation. Although Simon frequently acknowledges his sources in *L'Acacia* – photographs, pictures, documents, more or less eyewitness accounts, the elderly relatives questioned in Chapter 7 – he does not parade intertextual distance but rather uses them, however tentatively, to construct something plausible which might even be true. The sense of plausibility is increased by the days and dates of the chapter headings which root the narratives in specific historical time, and by the way in which the characters are framed in history. Eleven pages at the beginning of Chapter 3 are devoted to the history of an infantry regiment mobilised in 1914. Thereafter the captain emerges, almost as an example: 'Parmi ceux qui tombèrent dans le combat du 27 août se trouvait un capitaine de quarante ans dont le corps encore chaud dut être abandonné au pied de l'arbre auquel on l'avait adossé' (L'A, 61).[30] The novel often moves in this way, from general to particular, or between general and particular, each reinforcing the credibility of the other. The anonymity of the characters creates in part the same effect. By refusing to name the principal characters as himself, his father, mother and the members of his family, Simon declares their status as fictions, hypothetical constructs, and announces his suspicion of retrospective synthesis. The only syntheses he allows are those which name the characters by their function in particular social contexts. The mother is 'la jeune fille' or 'la veuve'; the father, 'le jeune boursier', 'le capitaine'; the son, 'l'enfant', 'le touriste', 'le peintre en cubisme', 'le réserviste', 'le brigadier'.[31] On the other hand, the characters' anonymity also signifies that Simon has refused to give them fictional names and it tends to generalise their significance; it transforms them from individuals into types, representative of French experience.

More than previous novels, *L'Acacia* emphasises shared French experience. The comparisons and contrasts between the two families are not arbitrary or singular. Simon exploits the material at his disposal to bring out the contrasting beliefs, values, traditions and modes of being of two of the constituent parts of the society of the Third Republic. In a sentence of Proustian balance and humour he sums up the passionate, heterogeneous beliefs of the captain's sisters:

> De leur éducation, de souvenirs de silhouettes titubantes le soir dans la rue du village, de cris de femmes battues, de javelles d'où tombait parfois en se tortillant une mince lanière couleur de fer, des années dans les écoles de montagne autour desquelles la neige ne fondait que pour faire place à la pluie, et de la perte d'une cousine dont elles se rappelaient le mouchoir taché de sang, les deux institutrices devaient conserver jusqu'à leur mort un sentiment presque superstitieux tenant à la fois de la crainte et d'une viscérale répulsion qui leur faisait mêler dans une même terreur (baissant la voix si par hasard il leur arrivait d'en parler, comme si les mots eux-mêmes étaient chargés d'un pouvoir maléfique et salissant, comme obscènes) l'ivrognerie, les vipères, la boue, les prêtres et la tuberculose. (L'A, 65–6)[32]

This family of mountain peasants illustrates, as in its previous version in *L'Herbe*, though less schematically, the values of the Third Republic and a typical if accelerated success story: from illiterate peasant through primary schoolteacher to higher education in just two generations. The other, bourgeois family differs from it in almost all respects: landowning rather than land-working, Catholic, sensual, in slow indolent decline from a glorious past. The bourgeois family boasts descent from a Napoleonic general; the peasant family proudly recalls a great-uncle who took to the woods to avoid being press-ganged into the armies of the 'ogre'.[33] Among the most delightfully humorous lines of the novel are those which recount the first incredulous visit by bourgeois mother and daughter to the prospective sisters-in-law, whom the old lady mistook for servants (L'A, 131–2).

Equally, the contrasts between 1914 and 1939–40 are generalised to illustrate the facts and moods of the times much more explicitly than, for example, in *La Route des Flandres*. In 1914 crowds cheer, the officers are confident, the slaughter is immediate and massive. In

1939 the army mobilises in demoralised resignation: '[O]n va tous au même endroit: au casse-pipe' (L'A, 167).[34] Discipline is slack; there is no ideological unity or sense of common purpose: when war is declared, the commanding officer can find no words to encourage his assembled troops. Isolated by the hostility of civilians and refugees, the cavalrymen mistrust their officers, as they do all those whose 'occult omnipotence' they blame for their situation. Simon's perspective is largely that of the ordinary soldier; he sees the war from the bottom up. From memory, he imaginatively reconstructs the feelings, prejudices and preoccupations of 1939-40, just as he uses documents and the family's oral tradition to do the same for the events of 1914 and before.

Simon's treatment of the past in *L'Acacia* can be illuminated by comparing it with another school of historical thought. Historicism, dominant in nineteeth-century Germany, reacted against the Enlightenment. It affirmed that human nature differed from civilisation to civilisation and that Reason alone could not reconstruct the human past. Hence the methods of the natural sciences were not appropriate to the study of human beings. These convictions, in direct conflict with the tenets of positivism, coincide largely with the negative assumptions underlying the way Simon writes *L'Acacia*: substantial knowledge of the past is not possible and no laws govern history. Furthermore, *L'Acacia* also shares some aspects of historicism's positive definition of history. Simon's concentration on individuals corresponds to a famous late nineteenth-century definition of the aim of the historical sciences: to depict 'human lives in the full richness of their unique development, perceived in their living individuality'.[35] Imaginative understanding, Dilthey's key requirement of the historian, is just what Simon brings to his study of the past. The parallels grow closer when one goes beyond these somewhat vague prescriptions to consider the relationship between truth and method in historicist thought. For historicism truth is not absolute but relative. Written history does not correspond to what really took place. Truth lies rather in coherence: this piece of evidence tallies with that. Added together, weighed and sifted, the evidence produces an overall pattern in which the whole confirms the parts and the parts the whole. This describes Simon's method in *L'Acacia*. The mother was religious: this coheres with the evidence that she went regularly to church. The father was shot in the forehead: though suspect as a cliché, the evidence of the witnesses coheres both with army tradition and the father's adoption of the values of the officer class.

L'Acacia even opens the door to further levels of coherence. In *La Route des Flandres* Georges declared that to know de Reixach's death was a fact, he would have had to be also in the position of the sniper. This statement implies a demand for absolute truth in the tradition of positivism. In the absence of this all-round perspective, Georges despairs of all knowledge. In *L'Acacia*, however, Simon provides an alternative perspective on the confusion of the battle lines on 17 May. He notes that a German general later related how the noise of battle made it impossible for him to get his men to stop firing even although his entire armoured column had already passed through and beyond the retreating French.[36] This account coheres with that of the corporal. Thus from two perspectives evidence accumulates about the murderous confusion of the Flanders road on 17 May 1940. In *L'Acacia* Simon moves towards the view that there are interpersonal standards by which truth can be judged. It is not therefore inappropriate to note that Simon's recital of abandoned fortifications and incompetent leadership tallies with other eyewitness accounts of the débâcle, for example Marc Bloch's *L'Etrange défaite*. *L'Acacia* implies assent to Bloch's affirmative appeal: 'Que chacun dise franchement ce qu'il a à dire; la vérité naîtra de ces sincerités convergentes.'[37]

There remains however a gulf between *L'Acacia* and any work of history. Simon is not a historian in the sense in which that word has come to be used since the Enlightenment. He is not primarily concerned with comprehensiveness of knowledge, vigour of argument, factual truth. His characters transcend both the singular and the typical. They are larger than life. They have the coded recurrent epithets of epic heroes: the 'regard de faïence' and 'barbe carrée' of the father, the 'mains crevassées' and 'visages ravinés' of the sisters.[38] Unnamed, they aspire to the universality of myth. Their fate, too, has a mythical quality. In blending his family's history with that of French society, Simon produces a myth of decline and destruction. The mood is set by the first chapter: the devastation of the battlefields marks the end of an era. Over the ambitions of the peasant family, the self-sacrifice of the sisters, the industry and dedication of the young brother, his successful entry into the officer caste, his attainment of the impossible prize, the sleeping beauty – over all this hangs the shadow of his early death. Education, the acquisition of skills and knowledge, social advancement, have served no purpose; these myths of progress are destroyed by the wall of fire which meets the regiment at the front in 1914, 'quelque chose qui ne ressemblait ni à une charge ni à rien de ce qu'ils avaient pu apprendre dans les livres ou sur le terrain' (L'A, 55).[39] A parallel fate overtakes the other family. Its illustrious name has already been lost. The dilettante 'député-poète' also perishes in 1914. The sole surviving adult male is the cousin, officer by influence, foolhardy, eccentric. The débâcle of 1940 is an epilogue to the destruction of this society. Here history certainly does repeat itself as farce. If the code of honour of the officer class, its exclusive aristocratic tone, its nonchalant disregard of danger, were tragically inappropriate in 1914, they survive in caricature in 1940. The corporal looks back with comic despair on his colonel's attempt to lead the three members of his regiment back to their own lines: 'pensant plus tard Mais pour le rôle qu'il nous destinait à jouer ça n'avait pas d'importance aussi bien j'aurais pu me trouver ficelé à califourchon sur un âne la tête tournée vers la queue' (L'A, 293).[40] As for the urban France of 1982, its fume-filled streets, garish shop windows and piped music are a footnote to history, the ultimate parody of progress.

These aspects of myth are familiar from earlier novels, though never so fully elaborated. What distinguishes *L'Acacia*, however, is a new myth of self. In earlier novels the self is sought but cannot be found, it disintegrates and disperses, its origins cannot be traced. *Les Géorgiques* begins a process of integration which *L'Acacia* carries much further. *L'Acacia* affirms the value of self. First, in relation to the family. Although both branches of the family decline, in coming together they serve a purpose. As Simon presents it, the idyllic marriage of the captain and the 'inaccessible princess', though so brutally cut short, had yet served an end: it had produced a son. This mission accomplished, the father's trajectory was over, he could go. Second, the value of self is affirmed through the new meaning which Simon gives to war in *L'Acacia*. As before, part of its effect is to undo the young soldiers who undergo it. The corporal becomes a creature of instinct; animal imagery runs through *L'Acacia* as through *La Route des Flandres*. But in *L'Acacia* that undoing has strongly positive connotations. Simon pours scorn on the pre-war existence of the corporal, his role-playing as cubist painter, anarchist, moneyed tourist. War, by contrast, awakens him. It puts him in touch with the elemental in nature. His perceptions, previously blinkered by convention, become fresh and acute. Simon considered calling this novel 'le lion noir'; the corporal was holding the lid of a tin of polish with a black lion on a red background, when the alert was given and action began on 10 May 1940. It would have been an appropriate title in that this moment and these bright colours and shapes symbolise the corporal's awakening to the vivid essentials of perceptual experience. In the final chapter, the first meaningful task which the corporal, returned from the war, sets himself is to draw just such elemental natural shapes with detailed precision. When, on the last page of the novel, he sits down at his desk with a blank sheet of paper, it becomes clear that *L'Acacia* is both the story and the product of a vocation. Indeed one of the pleasures and glories of the novel is its continual shifting of focus, from long shot to close-up, from grandiose to miniscule, from the great sweep of history to the arbitrary and precise. The war has turned the corporal into a writer, the writer whom we are now reading. Amongst other things, *L'Acacia* recounts the mythic origin of a corporal–writer who carries the Nobel Prize for literature in his knapsack.

L'Acacia, then, affirms the value of self and of writing. A possible link between these affirmations is to be found in Simon's reworking of family relationships.

The first chapter of *L'Acacia* recapitulates the family relationships of previous novels. It describes a vain search for the missing father. Simon sets up here, as elsewhere in his work, an analogy between public and personal: the devastation of the battlefields stands as an image for the trauma of personal loss, just as the personal loss confirms the collapse of civilisation. The absence of the father is indistinguishably both proof and source of a 'cataclysme définitif, total' (L'A, 15), the end and absence of history. But the father's place is not totally void since the child's mother decrees that the grave has been found, and makes him kneel before it. She takes the father's place in prescribing values to the child: patriotism, honour, fidelity, piety. The power of these values is potentially all the greater since the mother uses the dead father to sanction them, as if she were not the source but merely the transmitter of values. Yet, though wife and sisters are united in idolising the dead captain, the families are not ideologically homogeneous: when mother and child kneel at the graveside, the sisters remain standing. One might say that the child escapes the phallic mother by slipping through this ideological crack.

For *L'Acacia* goes much beyond recapitulating previous relationships. *Les Géorgiques* provides a good introduction to what follows this first chapter. It illustrates two contrasted ways of dealing with the past, symbolised in the mass of papers and documents which the family inherits from L.S.M. The grandmother is haunted by the past but, unable to come to terms with it, she keeps it hidden, sealed, papered over. Thus repressed, the past manifests itself as a disabling wound. The grandmother takes upon herself responsibility for 'la série de deuils qui avaient frappé la famille' (LesG, 182).[41] She does penance by regularly inviting to her house a cousin who pours scorn on the ancestor of whom she is both proud (she wears the family brooch) and ashamed: he had killed the king. One may choose to read here the projection of a child–father relationship: the child, incapacitated and guilt-ridden at the father's death, is horrified above all at the idea of patricide, figured in the killing of the king. Such a reading is reinforced if one considers a second, contrasting pattern in *Les Géorgiques*. The grandson–old man–writer who inherits L.S.M.'s papers brings them into the light of day, and thus assumes the task of coming to terms with the past. The reconstruction of L.S.M.'s life, based on documents and sympathetic imagination, is a model for the procedure Simon will later follow in *L'Acacia*. As in *L'Acacia*, Simon sets up parallels between the lives of ancestor and descendant. Both are soldiers, writers, lovers of horses and (at a distance) of the land, disillusioned idealists, old men whose time is running out. An intertextual reading of *La corde raide* and *Histoire* provides a further parallel: time has not eased the pain of losing 'cette femme adorée ensevelie dans le néant depuis si longtemps et dont le souvenir après vingt ans me déchire le coeur' (LesG, 76).[42] For the first time in Simon's work the ancestor–father becomes a living complex figure with whom the descendant–child sympathetically identifies, in his weaknesses and strengths. Above all, the descendant clarifies and works through what his grandmother could not face: L.S.M. as murderer. L.S.M. defied authority by killing the king; he did violence to his own family by killing his brother. The descendant understands and accepts what could be seen as two fragmented parts of patricide. Indeed he admires his ancestor's achievement, 'l'exploit titanesque d'accoucher un monde et de tuer un roi' (LesG, 149).[43]

Les Géorgiques, then, figures reconciliation with the past through a new positive relationship between the generations. In *L'Acacia* reconciliation is more complete and comes closer to home. Ancestor and descendant are replaced by father and son, whose lives are counterpointed in patterns of positive identification. Their contrasted early lives – the father's purposeful and determined, the son's aimless and floating – come together in the experience of war, in mobilisation, parting from loved ones and coming under fire. Then comes a difference, crucial to reconciliation, brought out by the arrangement of Chapters 2, 3 and 4. Chapter 2, '17 mai 1940', ends abruptly at the very moment when the enemy opens fire to massacre the cavalry squadron: ' "Faites passer: les blindés allemands sont dans le village! Arrêtez! Blindés dans le village! Faites passer! Les Al . . ." ' (L'A, 48).[44] This is when the son ought to have died. Chapter 4 takes up the son's story immediately after, or rather from the moment when, to his amazement, he discovered that he was not dead: 'On n'entend plus tirer. Accroupi maintenant, il regarde autour de lui' (L'A, 87).[45] That indefinite period in which the corporal was temporarily dead to the world ('il ne pourrait pas non plus dire combien de temps il est resté sans connaissance') is filled in the text by Chapter 3, '27 août 1914', which tells the story of the losses of 1914 and of the captain's death. Textually, then, the father dies in the son's place. Or, to put it another way, Simon here carries out the most tender of executions: the father dies and the son takes his place. From this point on the son will change and develop, eventually to find his vocation as writer.

The successful patricide of *L'Acacia* shows how far Simon has come since *La Route des Flandres*. In Chapter 6 (pp. 111–12) we studied the relationship between Georges and de Reixach and argued that it showed an unresolved Oedipus conflict. Georges realises part of the Oedipal fantasy by sleeping with Corinne; he fails to rid himself of the overbearing father-figure to whom he shows hostility and resentment; he fails to achieve a sense of personal identity. *L'Acacia*, by contrast, shows an Oedipal crisis resolved. Simon salutes the father, treats him with dignity and respect. He successfully constructs his death, and affirms the worth and autonomy of the son.

This argument should not be misunderstood. It does not concern Claude Simon, the man, husband, traveller, recipient of honorary degrees. It was not in his early seventies, in writing *L'Acacia* that Claude Simon came to maturity. Freudian theory, so divided as to the nature and timing of successive stages of human development, is quite clear that the traces of each stage coexist as fantasies in the adult psyche. The argument is rather that Simon, the novelist, has quarried these fantasies in a particular order. The changes concern every aspect of his works. All the novels up to *La Bataille de Pharsale* are variously presided over by the absent father and the all-too-present mother. The father-figures are sometimes prestigious but resented and undermined; the mother-figures loved and feared. The fantasies are Oedipal or pre-Oedipal. In parallel, history, like the father, is essentially absent. It becomes a mythical figure; personified, it attracts some of the same violent feelings shown towards father and mother: it is prestigious and denigrated, determining though non-existent, destructive and all-consuming. In the novels of the formalist period, from *Les Corps conducteurs* to *Leçon de choses*, these themes go underground. Or rather they are fragmented, churned and dissolved by language. They can be glimpsed but only as particles suspended in a liquefied mass. When the family and history re-emerge in *Les Géorgiques* and *L'Acacia*, they have a different colouring. Both novels show identification with the father, Oedipus resolved. That change goes hand in hand with a new view of history: it is not irremediable absence, the past can, tentatively, be constructed, nature renews itself. Both these sets of changes culminate in the new prestige which *L'Acacia* attributes to the self, apogee of the family line, justified by the vocation of writer.

One may ask whether the order in which Simon has deployed the themes of family and history has been in any sense determined. Could the order have been reversed? Could the works of Simon's maturity from *Le Vent* onwards have begun with the relative sense of completion and plenitude typical of *L'Acacia* and moved towards the feelings of anguish and loss which reach their highest pitch in *La Route de Flandres*? This seems unlikely, for two reasons. The first is historical. Simon's works are involved twice over in history. Their subject-matter is largely the great events of the first half of the twentieth century – the two world wars, the Spanish Civil War – with backward glances to the nineteenth century and the Revolution of 1789. Thematically, intellectually and formally, however, Simon's works are closely bound to the intellectual life of the latter part of the century. Broadly, they are concerned with the collapse of the ideologies of the left, that slow crumbling from the 1930s onwards which, by the end of the 1980s and the extinction of the Soviet Union, had left liberal capitalism in undisputed intellectual and physical mastery of the globe, for good and ill. More particularly, Simon has been sensitive to the intellectual trends of the times. His works have attracted the various critical approaches adopted in successive chapters of this book, because they have reflected and contributed to the fluctuations of intellectual debate. *La Route des Flandres* dramatises a crisis in representation; it shows how a certain use of language challenges a realist aesthetic based on fidelity to perception and memory. Here Simon, in his own way, took the linguistic turn which would manifest itself in the anti-humanist modes of thought and critical practice which dominated the 1960s and 1970s: structuralism; deconstruction; intertextuality (in its extreme form); Lacanian psycho-criticism (to the extent that, in style and substance, its practitioners imitate the master's suspicion of rationality and of the 'classic European overvaluation of the individual conscious mind').[46]

Simon's own practice from *La Bataille de Pharsale* to *Leçon de choses* had most affinity with the variant of structuralism promoted by Ricardou. At this stage Simon was clearly stimulated by a particular form of criticism, just as Simon's novels, among others, challenged Ricardou to systematise his critical theory. To some extent the supremacy which Ricardou established over the 'nouveau roman' and its critics through the Cerisy colloquia in the early 1970s was out of phase with the prevailing cultural climate. Ricardou was propounding his ultra-rationalist schema when the confident scientism of early structuralism was already past its heyday, the early to mid-1960s. Following the psycho-drama of 1968, Barthes's *S/Z* contributed to refocalising interest on desire and reader response. But if some of Simon's critics lingered almost in a time-warp in the 1970s, *Les Géorgiques* and *L'Acacia*, manifestations in part of a revolt against Ricardou, set Simon in tune with the 1980s. With the collapse of the ideologies of the nineteenth century began a reaffirmation of the values of the Enlightenment and some questioning of what were bracketed as the philosophies of the 1960s: there was discussion of 'the *return* of the subject, of the referent, of history'.[47] Autobiography, as practised by Simon, Sarraute and even Robbe-Grillet posits the writer as in some sense the proprietor of his own text; it makes claims for rationality and the control of the conscious mind. Coherent narrative, contoured characters, re-emerge in Sarraute, and even in Sollers,[48] as they do in Simon. Imagination and reason, working together, offer hope of some knowledge of the past; they imply the possibility of inter-subjective truth and an appeal to the principle of verification.

But finally we must turn to psychocritical theory to find confirmation that the road Simon builds stone by stone from *La Route des Flandres* to *L'Acacia* could not have been constructed in the reverse direction. The most consistent, most deeply grounded of Simon's themes is the search for origin. Psychoanalytic theory relates that search, not unusual in literature, to the infant's desire for power. Marthe Robert, analysing *Don Quixote*, remarks that 'l'avènement de Don Quichotte révèle la toute-puissance du désir (et le désir de toute-puissance) qui est le génie propre de la pensée infantile.' A characteristic form of this desire is to insist on being responsible for one's own origin: 'l'homme tombé du ciel [. . .] prétend s'engendrer lui-même sans le concours de parents humains (ce qui ne l'empêche pas de se passionner pour les familles fictives dont ses livres de chevalerie racontent les hauts faits, c'est pourquoi tout en reniant ses ascendants, il a le goût des généalogies)'.[49] Simon is less radical, less fantastic than Cervantes: the family which he gradually constructs is a modified, mythified version of his own. Beginning with fragmentary fictions and families, he gradually adjusts family and fictions to make them cohere. In *Histoire* he works on the mother's household; the father remains in shadow. In *Les Géorgiques* a fuller lineage on the mother's side is established and the various fictions – of childhood, Spain and the Second World War – are attributed to a single source. In *L'Acacia* the two sides of the family are at last brought together, and the father, cast previously in so many partial and rejected moulds, is now given pride of place. The serenity of *L'Acacia* springs ultimately from the writer's confidence that he has replaced his family tree with a tree of his own devising: his parents, relatives and ancestors have become his children. The last few lines of *L'Acacia* repeat in modified form the first few lines of *Histoire*. These modifications are in themselves a statement of power: what I have made, I can unmake and remake. But more than that, Simon quotes *Histoire* to point back at the written self which has replaced the contingent human being. The integration achieved in *L'Acacia* is provisional, subject to revision in Simon's next novel. But what *L'Acacia* celebrates is unequivocally the construction of a life and a life's work. The sense of the novel lies in that final, modestly triumphant gesture towards *Histoire*: 'l'intertextualité', as Simon might have put it, 'c'est moi'.

Notes

Introduction

1 'When I mentioned my various adventures a moment ago (revolution, war, escape, illness, travels, etc.) I forgot to include what in my view is the main one. I have written books. And, believe me, that's some adventure!' A. Poirson and J. P. Goux, 'Un homme traversé par le travail. Entretien avec Claude Simon', *La Nouvelle Critique*, 105, June–July 1977, p. 39.

2 The following books have appeared in English: C. Britton, *Claude Simon: Writing the Visible*, 1987; M. Evans, *Claude Simon and the Transgressions of Modern Art*, 1988; R. Sarkonak, *Understanding Claude Simon*, 1990; M. Orr, *Claude Simon: the Intertextual Dimension*, 1993; C. Britton (ed.), *Claude Simon*, 1993.

3 In *Album d'un amateur*, 1988, Simon wrote a number of short texts to accompany his own photographs in black and white, pre-war, and colour, post-war. *Photographies 1937–1970*, 1992, was published to coincide with exhibitions of Simon's photographs held in Paris and Toulouse. It includes a preface by D. Roche and a short note by Simon.

4 This aspect of Simon's work forms one of the main themes of Stuart Sykes's introductory study, *Les Romans de Claude Simon*, 1979.

5 J. Chalon, 'Enquête. Les débuts obscurs d'écrivains célèbres', *Le Figaro*, 11 March 1972, p. 14.

6 *Le Tricheur* and *La Corde raide* were originally published by Sagittaire. After Sagittaire went bankrupt, Simon published *Gulliver* and *Le Sacre du printemps* with Calmann-Lévy. Meanwhile Jérôme Lindon had bought the rights to Sagittaire's publications. After the appearance of *Le Sacre du printemps* he wrote to Simon inviting him to come to Minuit on the basis of the original contract with Sagittaire. Calmann-Lévy objected and took the matter to court, but lost their case. See 'A

l'intention des auteurs', *France-Observateur*, 13 February 1958, p. 18.

7 Many were later reworked and published in *Pour un nouveau roman*, 1963. Some of the letters were reprinted in F. Jost (ed.), 'Robbe-Grillet', *Obliques*, 16–17, 1978.

8 'Le réalisme, la psychologie et l'avenir du roman', *Critique*, 111–12, 1956, pp. 695–701.

9 E. Henriot, 'Un nouveau roman', *Le Monde*, 22 May 1957.

10 See 'Les raisons de l'ensemble', *Conséquences*, 5, 1985, pp. 62–77.

11 See especially 'La littérature comme critique', in *Pour une théorie du nouveau roman*, 1971, pp. 9–32.

12 *Degrés*, 1960, was published by Gallimard. Butor stayed with Minuit for his four volumes of critical essays, *Répertoire*, of which the first was also published in 1960.

13 See 'Terrorisme, théorie', in J. Ricardon (ed.), *Robbe-Grillet: analyse, thé orie*, vol. 1, 1976, pp. 11–63.

14 See L. Oppenheim (ed.), *Three Decades of the French New Novel*, 1986.

15 'rejection of a certain academic, conventional kind of literature, rejection of all plot, agreement – at least between Alain Robbe-Grillet and myself – on the non-significance of the world', M. Chapsal, 'Le jeune roman', *L'Express*, 12 January 1961, p. 31.

16 'People who write generally imagine that everything has to be said, or rather that there must be no holes. So they replace the moments of absence, which exist in reality, moments when they have felt nothing, perceived nothing at all, by a sort of greyish cement which is meant to bind and which seems to me very false.' M. Chapsal, 'Entretien avec Claude Simon', *L'Express*, 10 November 1960, p. 30.

17 Ainsi un roman est-il pour nous moins *l'écriture d'une aventure que l'aventure d'une écriture.*' *Problèmes du nouveau roman*, 1967, p. 111 (first published 1961).

18 *Qu'est-ce que la littérature?*, 1948, p. 29.

19 'It submerges us. We organise it. It falls to pieces.
We organise it again and ourselves fall to pieces.'

Chapter 1

1 'a course made of loops which form a clover leaf, such as can be drawn by hand without ever raising the pen from the surface of the paper', 'La fiction mot à mot', in *Nouveau roman: hier, aujourd'hui*, vol. 2, 1972, p. 89.

2 'And while the solicitor was talking to me, was relaunching himself – for perhaps the tenth time – on this story (or at least what he knew of it, or at least what he imagined of it, since of the events which had been unfolding over the last seven months he possessed, like everyone else,

like their heroes, their actors, only that fragmentary incomplete knowledge, composed of a sum of fleeting images, partially apprehended, of words, imperfectly grasped, of sensations, imprecisely defined, and all of it vague, full of holes, of gaps which imagination and an approximate logic strove to remedy with a series of risky deductions . . . and now, now that it's all over, to attempt to report, to reconstitute what happened, is a little like trying to re–glue the scattered incomplete fragments of a mirror, making clumsy efforts to fit them together again, getting a result which is incoherent, derisory, senseless.'

3 'Attempt to reconstitute a baroque altarpiece.'

4 'And as he went on recounting the scene to me, it seemed to me that I was living it better than he did, or at least was able to reconstitute a schema corresponding if not to what had really happened then in any case to our incorrigible need of reason.'

5 'as if', 'a little as if'; 'not . . . but', 'if not . . . then', if not . . . at least'.

6 'Even claiming that he (or rather his father acting in his stead, for no one would concede that he had even this kind of ability) had used the final remnants of prestige – the sabres, the weapons inlaid with precious metals, the ancestral portraits, the aristocratic name, the dusty old town-house – to ravish the consent, not of the principal interested party, the young woman, but of the parents (former gardeners, it was said, who in the space of a lifetime had used their hands to wrench from the earth approximately the equivalent of what the dashing generals, their descendants and and their stewards had taken a hundred years to squander on gambling debts, on horses, on women, or, with even greater certainty, on business deals too good to be true); his father, then, setting up a couple.'

7 'that insipid arrangement, that cement good only for filling holes'.

8 For example, D. Carroll writing on *Le Vent* in *The Subject in Question: the Languages of Theory and the Strategies of Fiction*, 1982; and J. Duffy in 'Antithesis in Simon's *Le Vent*: authorial red herrings versus readerly strategies', *Modern Language Review*, 83, 1988, pp. 571–85.

9 Montès shares Prince Myshkin's saintly, comic innocence, his unfathomable character (is he stupid or intelligent?), his unhierarchised perceptions. The first words of the novel, 'Un idiot', can be read as Simon's homage to Dostoevski. Various parallels between Simon and Faulkner, paticularly strong in the period between *Le Vent* and *La Route des Flandres*, have been traced by A. Duncan, 'Claude Simon and William Faulkner', *Forum for Modern Language Studies*, 9, 1973, pp. 235–52; S. Sykes, 'The novel as conjuration: *Absalom, Absalom!* and *La Route des Flandres*' and J. de Labriolle, 'De Faulkner à Claude Simon', both in *La Revue de littérature comparée*, 53, 1979, pp. 348–57 and 358–88.

10 'No one makes history, it can't be seen, any more than grass can be seen

growing.'

11 'the same face which now, mummified (ossified), rested on the pillow, identical, imprinted with that same invincible expression of peaceful virginity'.

12 'not like a slice of time, precise, measurable and limited, but as a vague duration, chopped up, composed of a succession, an alternation of gaps, of dark spaces and light spaces'.

13 'to endure history (not to resign onself to it: to endure it), is to make it'.

14 The techniques of this creative momentum have been studied in detail by G. Roubichou. See in particular *Lecture de 'L'Herbe' de Claude Simon*, 1976, pp. 137–213.

15 'He looked at me than at the letter then at me again.'

16 'Le cheval', *Les Lettres nouvelles*, February and March 1958, pp. 168–89, 379–93.

17 'And she now no longer invented (as Blum used to say – or rather fabricated during the long months of war, of captivity, of enforced continence, from a brief, single sighting on the day of a horse show, from Sabine's tittle-tattle or from the scraps of sentences (representing scraps of reality), of confidences or rather near-monosyllabic grunts wrested by patience and cunning from Iglésia, or from even less: from an engraving which didn't even exist, from a portrait painted 150 years before . . .), but such as he could see her standing now, really, truly, before him, since he could (was about to) touch her.'

18 'did I really see him or believe I saw him or imagine it afterwards or even dream it'.

19 'in the midst of this kind of decomposition of everything as if not an army but the world as a whole and not just simply in its physical reality but even in the representation which the mind can make of it . . . was being undone, was falling apart, was disappearing into pieces, into water, into nothing'.

20 Simon spoke of 'la mort de Georges, mort symbolique', in an interview with H. Juin, 'Les secrets d'un romancier', *Les lettres françaises*, 6–12 October 1960, p. 5.

21 'bad show'.

22 'and he: Reichac good God not got it yet: chac the x like sh and the ch at the end like k Cor what a dope he is I've told him at least ten times never been to the races oaf it's a well known name all the same'.

23 'And Georges (unless it was still Blum, interrupting himself, playing the fool . . . and Blum (or Georges): "Have you finished?", and Georges (or Blum): "I could go on", and Blum (or Georges): "Well go on." '

24 ' "Ah yes! . . ." said Blum (now we were lying in the dark, that's to say so intertwined piled on top of one another that we couldn't move an arm or a leg.'

25 'listening to the silence, the dark, the peace, the imperceptible breathing

of a woman beside him'.

26 'now we were lying'; 'now all three of them were standing'; 'being able now to hear the air entering her'.

27 'imbued, filled with a superstitious confidence in those words which he could only understand and pronounce (and even then perhaps making a mess of them)'.

28 'knowledge, learning, what books enclose'.

29 'because I would like never to have read a book, never to have touched a book in my life, not even to know that there exists something called books, and even, if possible, not even to know, in other words to have learned, in other words to have allowed myself to be taught, to have been stupid enough to have believed those who taught me that characters laid out in lines on white paper could signify anything other than characters on white paper, in other words precisely nothing'.

30 See the interview with Simon by G. le Clec'k, 'Claude Simon, prix de la nouvelle vague: "je ne suis pas un homme orchestré"', *Témoignage chrétien*, 16 December 1960, p. 19.

31 'if the contents of the thousands of books of that irreplaceable library had precisely been incapable of preventing things like the bombardment which destroyed it, I couldn't clearly see what loss for humanity might be represented by the destruction under phosphorus bombs of these thousands of books and papers which were manifestly devoid of the slightest utility. There followed a detailed list of things of certain value, necessities of which we have much greater need here than of the entire contents of the celebrated library of Leipzig, namely: socks, underpants, woollens, soap, cigarettes, sausage, chocolate, sugar, tinned foods, dou. . . .' The three letters which begin the last, uncompleted word in this list leave scope for the translator's imagination: 'gal . . .' could be the beginning of 'galoche' (clog), 'galetouse' (mess-tin) or 'galette' (military slang for hard biscuit or money). What word did Richard Howard have in mind when he translated 'gal . . . ' as 'ma . . . '(Calder, 1985, p. 176)?

32 'I wasn't a man any more but an animal a dog more than a man a beast.' For a perceptive study of animal imagery in *La Route des Flandres*, see K. S. Brosman, 'Man's animal condition in *La Route des Flandres*', *Symposium*, 29, 1975, pp. 57–68.

33 'this reckless abundance, without beginning, or end, or order'.

34 A detailed study of the semantic and morphological generation of *La Route des Flandres* and the rhythmic function of punctuation has been made by D. Lanceraux in 'Modalités de la narration dans *La Route des Flandres*', *Poétique*, 14, 1973, pp. 235–49.

35 'a bit taken aback, impatient as if in a drawing-room someone had suddenly accosted him without having been introduced or interrupted him in the middle of a sentence with some inopportune remark (like for

example'.

36 'renewing with the little sub-lieutenant his peaceful conversation of the sort held by two cavalrymen riding together (at training or at the racecourse) and which no doubt ran to horses, to fellow officers, to hunting and racing. And I seemed to be there, to see it: the green shade with women in coloured print dresses.'

37 'And this time Georges could see them, just as if he had been there himself.'

38 'In short: perhaps he thought, as one might say, that he could get two birds with one stone, and that if he managed to mount the one he would tame the other, or vice versa, in other words that if he tamed the one he would as victoriously mount the other, in other words that he would get her to the finishing post, in other words that his finishing post would get her victoriously to where he had doubtless never managed to lead her.'

39 'able to see, as if he'd been no more than a few metres away, the neck of the filly covered with grey foam where the reins dangled, the group, the hieratic medieval cortège still proceeding towards the brick wall, having now traversed the crossing of the figure of eight, the horses hidden again up to their bellies by the enclosing fences half disappearing so that their bodies seemed cut in half only the upper part rising above seming to glide on the field of unripe wheat like ducks on the still surface of a pond I could see them progressively as they turned to the right into the sunken road with him at the head of the column'.

40 'like ducks with their heads chopped off which go on walking fleeing grotesquely covering a few metres before collapsing for good: nothing but a story of slit throats all in all since according to the tradition the version the flattering family legend it was to avoid the guillotine that the other one had done it had been constrained to do it So they shold have changed their crest from that day on replaced the three doves with a headless duck'.

41 'then I flung myself to the ground dying of hunger thinking Horses eat it why not me I tried to imagine to persuade myself I was a horse, I was lying dead at the bottom of the ditch devoured by ants my whole body changing slowly through myriad tiny mutations into insensible matter and then it would be the grass that would feed on me my flesh fertilising the earth and after all not much would have changed, except that I would simply be on the other side of its surface just as you can go through a looking-glass where (on that other side) things would go on happening symmetrically,in other words above it would go on growing still indifferent and green just as, they say, hair goes on growing on the skulls of the dead the only difference being that I'd be pushing up the daisies gobbling up her pussy sweating our beaded bodies giving off that bitter strong smell of root, of mandrake, I had read that shipwrecked men and hermits fed on roots on acorns and then she took it

first between her lips then altogether into her mouth like a greedy child it was as if we were drinking each other, quenching each other's thirst gorging satisfying ourselves ravenous hoping to appease to still my hunger a little I tried to chew it thinking It's like salad'.

42 Literally: 'I'd be gobbling dandelions' ('pisse-en-lit' means 'piss-in-bed') to 'gobbling where she pisses'.

43 'it seemed to have renounced the spectacle of this world to turn its look the other way, to concentrate it on an interior vision more restful than the incessant bustle of life, a reality more real than the real'.

44 'so as to . . . summon up the sparkling luminous images by means of the fleeting incantatory magic of language, of words invented in the hope of making palatable – like those vaguely sweetened mixtures in which bitter medecines are disguised for children – unspeakable reality'.

45 'like a parodic appendix to the battle'; 'nothing is worse than silence'. S. Sykes studied this aspect of the novel in 'The Novel as Conjuration: *Absalom, Absalom!* and *La Route des Flandres*', *La revue de littérature comparée*, 53, 1979, pp. 348–57.

46 'In a strange, final reversal, this world consumed by disaster appears in the end richer than all the forms which destroy it. Death reigns throughout; but it cannot prevent another power – which one can only call life – from having the last word. . . . It is rather the profusion, the lyricism [of this world] which overwhelms us, as if, freed from believing that we exist, we discovered, in its destruction, the inexplicable and jubilant confusion of existence.' 'Sur la route des Flandres', *Les Temps modernes*, 178, 1961, p. 1034.

47 'substantial knowledge, disenchanted and amazed'.

48 In *The Subject in Question: the Languages of Theory and the Strategies of Fiction*, 1982, p. 105, David Carroll remarks: 'In fact, as Lukács himself admits, the theory of realism is an essentialist doctrine; it consists in finding within what usually passes for realism the essential core of the real.' *La Route des Flandres* dramatises Simon's abandonment of realism through its failed search to find such a core.

Chapter 2

1 *Nouveau roman: hier, aujourd'hui*, ed. with F. van Rossum-Guyon, 2 vols., 1972; *Claude Simon: colloque de Cerisy*, 1975.

2 'Je ne me sens pas capable de faire de la critique de la critique. . . . Mais je veux faire savoir combien tous les textes d'analyse de Jean Ricardou m'ont intéressé, au point même que je cite souvent dans des conférences ou les interviews les pages écrites sur 'l'objet scriptural', pages absolument remarquables, qui m'ont personnellement beaucoup éclairé sur les mécanismes de la création littéraire et mes propres mécanismes.' 'I don't feel able to criticise criticism. . . . But I want to

make it known how much all the analyses of Jean Ricardou have interested me, even to the extent that in lectures or interviews I often cite his pages on "the written object", remarkable pages which have much enlightened me personally about the mechanisms of literary creation in general and my own in particular.' Claude Simon in '*Problèmes du nouveau roman*: trois avis autorisés', *Les Lettres françaises*, 11–17 October 1967, p. 13. See also similar remarks on the occasion of the appearance of Ricardou's *Pour une théorie du nouveau roman* ('L'opinion des romanciers', *La Quinzaine littéraire*, 121, 1–15 July 1971, p. 10), and the general tenor of 'Claude Simon, à la question', in *Claude Simon: colloque de Cerisy*, 1975, pp. 403–31.

3 *Critique*, 163, 1960, pp. 1011–24.

4 See *Nouveaux problèmes du roman*, 1978, pp. 9–11.

5 'With the New Novel, narrative goes on trial: it is simultaneously put in motion and in question.'

6 Ricardou defines and explains the concept of *mise en abyme* in *Le Nouveau roman*, 1973, pp. 47–50. The classic exposition of the *mise en abyme* and its uses is L. Dällenbach's *Le Récit spéculaire: essai sur la mise en abyme*, 1977.

7 *The French New Novel*, 1969, p. 26.

8 'Claude Simon and representation', translated by Annwyl Williams, in *Claude Simon*, ed. C. Britton, 1993, p. 73. (Originally published as 'Claude Simon et la représentation', *Critique*, 187, 1962, pp. 1009–32. (p. 1026).)

9 'Cinq notes sur Claude Simon', *Médiations*, 4, 1961–62, p. 6.

10 'Claude Simon, Merleau-Ponty and perception', *French Studies*, 46, 1992, especially pp. 39, 44, 45, 47.

11 R. Jakobson, 'Two Aspects of Language and Two Types of Aphasic Disturbance', *Language in Literature*, 1987, pp. 95–120. First published in *Fundamentals of Language*, 1956.

12 Genette's allegiance to a structuralist poetics can be traced in the essays collected in his three-volume *Figures*, 1966, 1969 and 1972; Todorov's structuralist phase is summed up in his *Qu'est-ce que le structuralisme? 2. Poétique*, 1968.

13 See, for example, the first section of 'Claude Simon, textuellement' ('Aspects de l'idéologie dominante'), in *Claude Simon: colloque de Cerisy*, 1975, pp. 7–10, and section G of 'La littérature comme critique' ('Le dogme de l'expression représentation'), in *Pour une théorie du nouveau roman*, 1971, pp. 20–2.

14 Ricardou does discuss these issues in various places, for example 'La littérature comme critique', in *Pour une théorie du nouveau roman*, 1971 pp. 9–32, 'Terrorisme, théorie', in *Robbe Grillet: analyse, théorie*, vol. 1, 1976, pp. 11–63, and, rather defensively, in 'Fonctionnement politique du texte', in *Nouveaux problèmes du*

roman, 1978, pp. 53–67, where he argues for an analogy between the hierarchical structure of some texts and the the hierarchical structure of society.

15 Althusser's *Pour Marx* was published in 1966. the *Tel Quel* group's development of literary theory, based partly on Althusser's revised view of Marx, was given fullest expression in *Théorie d'ensemble*, 1968. For a critical analysis of these relationships, see C. Britton, *The Nouveau Roman*, 1992, Chapter 4.

16 A flavour of the clash of ideas, styles and personalities between Ricardou and Robbe-Grillet can be obtained from Ricardou's opening paper at the Robbe-Grillet conference and the discussion which followed it: 'Terrorisme, théorie', in *Robbe-Grillet: analyse, théorie, 1. Roman/cinéma*, 1976, pp. 11–63.

17 See, for example, the paper by L. S. Roudiez and the discussion which followed it in *Claude Simon: colloque de Cerisy*, 1975, pp. 39–72.

18 'Yellow and then black for the blink of an eye and then yellow again: wings outspread like a crossbow quick between sun and eye darkness a moment on the face like velvet a hand a moment dark then light or rather recollection (warning?) recall of darkness flashing upward at lightning speed palpable, in other words successively chin, mouth, nose, forehead feeling it and even smelling it musty odour of vault of tomb like a handful of black earth hearing simultaneously the sound of torn silk the rustle of air or perhaps not heard perceived only imagined bird arrow beating whipping already gone feathers vibrating the deadly shafts criss-crossing forming a hissing vault as in that painting seen where? naval battle between Venitians and Genoese on a tossing spiky sea and from one galley to another the feathered humming arch in the dark sky one of them piercing his open mouth as he charged forward sword raised sweeping his soldiers with him transpiercing him silencing the cry deep in his throat

Dark dove haloed in saffron

In the stained-glass window white on the contrary wings outspread suspended in the centre of a triangle surrounded by diverging rays of gold. Soul of the Just man taking its flight. At other times an eye in the middle. In an equilateral triangle the perpendiculars, the bisectors and the medians intersect at the same point. Trinity, and she impregnated by the Holy Spirit. Vessel of ivory, Tower of silence, Rose of Canaan, Thingummy of Solomon. Or painted on the bottom of the inside as in the window of that pottery shop, opened wide. Who on earth buys things like that? Night vessel to receive. Squattings. Riddle: what is slitted, oval, moist and surrounded by hair? So an eye for an eye as they say a tooth for a tooth, or face to face. One looking at the other. Spurting thick and fast with a liquid hiss, like a horse. Or rather mare.

Vanished above the rooftops.'

19 'ACHILLES IMMOBILE AT FULL STRIDE. Zeno! Cruel Zeno! Zeno of Elea! / Have you pierced me with that winged arrow / Which vibrates, flies and does not fly! / The sound gives birth to me and the arrow kills me! / Oh! the sun. . . . What tortoise shadow / For the soul, Achilles immobile at full stride!'

20 'yellow becomes one of the demands to which *The Battle of Pharsalus* must submit'; 'Zeno, Achilles determine the demand for Greece, satisfied by Pharsalus: battle and O's visit'; '[T]he generator conflict, for example, certainly produces Pharsalus and other battles (from those of the Middle Ages to the battle of *The Flanders Road*, and including Kynos Kephalia), but also football match (near Pharsalus, indeed), and street demonstration, children's squabble and coitus (amongst other things by the traditional assimilation of feathered arrow with penis . . .)'. The quotations in this paragraph are from 'La bataille de la phrase', in *Pour une théorie du nouveau roman*, 1971, pp. 129, 127, 133.

21 'he who takes up the pen cannot be a writer preparing to represent what he sees or express what he feels. O, stripped of identity by the work of the text, caught up in the woof, product of his product, is a scriptor.' *Pour une théorie du nouveau roman*, 1971, p. 155.

22 'one must imagine the system as a whole as a mobile, changing shape round a very few fixed points'.

23 'Nothing other than a few words a few signs with no material consistence as if traced on air gathered conserved recopied traversing the colourless layers of time of the centuries at lightning speed rising from the depths and bursting at the surface like empty bubbles.' Simon juxtaposes here elements of a sequence which Sartre presents successively in *Les Mots*: words as objects, words as vessels of essential meanings, words as impotent gestures.

24 'I considered attentively in my mind some image which had obliged me to look at it, a cloud, a triangle, a steeple, a flower, a stone, sensing that beneath these signs there was perhaps something quite different which I should strive to discover, a thought which they were translating like these hieroglyphic characters which might simply be taken as representations of material objects.'

25 'the well-trodden beach was now no more than a confused muddle of hoof-prints superimposed on one another destroying one another'.

26 'Outside the window, meadows, woods, hills, drift slowly past. Vision is sometimes blocked or disrupted by the fleeting passage of hedges or trees which border the line. The head of one of the Spaniards, leaning forward, stands out against the bright changing background of the countryside swept sideways. The profile against the light is clear-cut, the nose hooked. The hair oiled and slicked back. The eyes fixed on the bulky Spaniard sitting on the facing bench, dressed in a brown shirt and

talking continuously, the other two – the thin one in profile and the bald one – restricting themselves to nodding whenever the fat one finishes a sentence with the word No pronounced with an interrogative inflection. However this question form which recurs at frequent intervals (every two or three sentences) seems to be a linguistic mannerism and to call for nothing but the imperceptible nods which it provokes.'

27 In the later 1960s Ricardou dropped the words 'creation' and 'creative' and substituted 'productive' and 'production' since he felt his earlier vocabulary was tainted by idealism and pandered to the myth of the artist creating from nothing (conversation with Jean Ricardou, 29 May 1980).

28 'When, for example in a precise description, the novelist establishes (and as we have seen, isolates) one of the qualities of an object (a triangle for example) this *independent* form provokes laterally a whole series of objects able to incorporate it (triangular). The same phenomenon occurs with each of the qualities of the originating object and induces corresponding series. . . . Several times suggested to the writer, these objects end up by imposing themselves. . . . By means of these controlled relays and chains . . . which are happening at every moment and at every level, the material of the novel is entirely invented by the exercise of description.' 'Plume et caméra', in *Problèmes du nouveau roman*, 1967, p. 72–3. Two other articles devoted to aspects of description, both first published in 1961, were 'La description créatrice: une course contre le sens', and 'Une description trahie', in *Problèmes du nouveau roman*, 1967, pp. 91–111 and 112–24.

29 'In a study of Ricardou's I found approximately this idea: when we perceive an object, the whole world is denied, cancelled. When I look at this glass it deprives me of all that I don't see. If I set out to describe this glass – which hides everything as I look at it – I have to enumerate the qualities which are common to this glass and to many other objects. Hence the glass is reattached to the world which reconstitutes itself in language.' J. Duranteau, 'Claude Simon. "Le roman se fait, je le fais, et il me fait"', *Les Lettres françaises*, 13-19 April 1967. Simon's view that perception denies the rest of the world bears echoes of Sartre's view of perception in *L'Imaginaire*.

30 'Starting from the axle which links the two large iron wheels, a chain (like a bicycle chain but bigger) goes horizontally to the right where it passes round a notched wheel with its inner portions cut away, like the crank wheel on a bicycle, after which it climbs to the left, curving under a pulley wheel, then rising again to a second toothed wheel, smaller than the first, then it drops, passes round three flat surfaces of different sizes and sets off to the right, back to the driving axle which also, by an arrangement of bevelled sprocket wheels, controls synchronously another chain, much stronger than the first, which disappears by rising

at a slant inside the machine in a plane parallel to the axle and hence perpendicular to the plane in which the wheels turn and the first chain moves.'

31 'the sales representatives remaining for a while yet, filing away their purchase orders and bills of exchange signed on village café tables, downing with a wry smile a last glass of local wine, leaving the café owner by way of artistic souvenir some sprucely coloured advertising panel adorned with machines and pretty girls, and eventually heading back for their Elysian fields of pastel-coloured skyscrapers, cancer-infested bricks, factories and giant airports'.

32 'MACCORMICK . . . a name which, through repetition and frequent appearance on this kind of machine, has been emptied of all Scottish resonance . . . and has become . . . with its dry sound, its metallic clank, something like the generic surname of everything which crawls on the surface of the earth with a creaking of chain and clashing iron.'

33 'Jalonnant de ses divers états tout le cours du récit, la moissonneuse-lieuse nous rappelle opportunément, en son insistance, que moisson et liaison, génération et transition, ne forment pas deux opérations disjointes.' 'Marking out the whole course of the narrative by its appearances in various states, the reaper–binder opportunely recalls, by its insistence, that reaping and binding, generation and transition, do not form two discrete operations.' 'La bataille de la phrase', in *Pour une théorie du nouveau roman*, 1971, p. 137.

34 *The Subject in Question: the Languages of Theory and the Strateties of Fiction*, 1982, p. 185.

35 See F. Van Rossum-Guyon, 'De Claude Simon à Proust: un exemple d'intertextualité', *Les Lettres nouvelles*, September 1972, pp. 107–37 (p. 115); R. Brox Birn and V. Budig, 'Deux hommes et un texte: Simon face à Rousseau, Proust et Orwell', *La Revue des sciences humaines*, 94, (220), 1990, pp. 63–78 (p. 69). '*La Bataille de Pharsale* represents Simon's reevaluation and subsequent rejection of the novel of Proustian quest in favour of one built upon detailed descriptions of surface lines, geometric forms, and colors', R. Birn, 'Proust, Claude Simon and the Art of the Novel', *Papers on Language and Literature*, 1977, pp 168–86 (p. 170).

36 'no doubt able, with a metallic clanking of elytrons, joints and mandibles, to carry out a series of simultaneous operations, like those insects which at one and the same time can move their multiple legs, flex their necks to right and left, open and reclose their claws, everything stopping suddenly, everything at once (in other words the insect no longer advancing, no longer moving head and claws) at the approach of a danger or at sight of a possible prey and then the silence, the terrifying immobility which precedes the sprint, the dash'.

37 'little by little they [these smashed pieces of machinery] give glimpses of

their incomprehensible, delicate, feminine anatomies, with their connecting parts, also delicate and complicated. Their joints, formerly oiled, rubbing gently against one another, are now jammed, stiff. Skywards, in emphatic and unending protest, vaguely ridiculous, like old divas, like old broken-down tarts, thay raise their skinny limbs.'

38 'The description (the composition) can be continued (or completed) almost indefinitely depending on the attention to detail with which it is executed, the force of the proposed metaphors, the addition of other objects visible in their entirety or fragmented by attrition, or time, or shock (or else because they appear only in part in the frame of the picture), without counting the various hypotheses which the spectacle can provoke.'

39 'the internal dynamics of writing'. 'Réponses de Claude Simon à quelques questions écrites de Ludovic Janvier', *Entretiens*, 31, 1972, p. 17.

40 See Ricardou's review of *Triptyque*, 'Un tour d'écrou textuel', *Le Magazine littéraire*, 74, 1973, pp. 32–3, and 'Claude Simon, textuellement', *Claude Simon: colloque de Cerisy*, 1975, pp. 18–19.

41 'I think one can get to everything by starting with the description of a pencil. The written world is not the perceivd world. But through language one makes discoveries: what's more, one "discovers" oneself in writing, in every sense of the word, and that's a risk which has to be taken.' J. Duranteau, 'Claude Simon. "Le roman se fait, je le fais, et il me fait" ', *Les Lettres françaises*, 13–19 April 1967, p. 4.

42 'That's the most difficult thing to conceive of: the transformation'; 'What's not being thought about here is the dialectic between these two processes.' These exchanges between Raymond Jean, Ricardou and Irène Tschinka are to be found in *Claude Simon: colloque de Cerisy*, 1975, pp. 267–8.

Chapter 3

1 In this category, despite varying methods and perspectives, one might include, for example, both J. A. E. Loubère in *The Novels of Claude Simon*, 1975, and S. Sykes in *Les Romans de Claude Simon*, 1979. The phrase 'centred wholes' was used by Simon himself and is to be found in *Nouveau Roman: hier, aujourd'hui, 2, Pratiques*, 1972, p. 92.

2 See, for example, C. H. Gosselin in 'Voices of the past in Claude Simon's *La Bataille de Pharsale*', *New York Literary Forum*, 2, 1978, pp. 23–33. Others moving in the same direction were: R. Mortier, 'Discontinu et rupture dans *La Bataille de Pharsale*', *Degrès*, 1, (2) , 1973, pp. c1-c6; F. Jost, 'Les aventures du lecteur', *Poétique*, 29, 1977, pp. 77–89; C. Gaudin, 'Niveau de lisibilité dans *Leçon de choses* de Claude Simon', *Romanic Review*, 68, 1977, pp.178–96. Umberto Eco

launched the word 'open' on a new career in *L'Oeuvre ouverte*, 1965; 'writeable' is an English version of Roland Barthes's coining, 'scriptible', in *S/Z*, 1970.

3 In *S/Z*, Barthes analyses Balzac's *Sarrasine* by fragments, in order to 'apprécier de quel pluriel il est fait'; the ideal 'writeable' text, however, is more irreducibly polysemic, therefore more fragmentary. Julia Kristeva introduced the notion of intertextuality in 'Bakhtine, le mot, le dialogue et le roman', *Critique*, 239, 1967, pp. 438–65. The reference to Derrida is a reformulation of Barbara Johnson's definition of deconstruction: 'the careful teasing out of warring forces of signification within the text', in *The Critical Difference: Essays on the Contemporary Rhetoric of Reading*, 1981, p. 5.

4 '*it seems that nature is reproduced here pell-mell in the ordr or rather absence of order in which it presents itself . . . for the German artist everything in nature is on the same plane the detail always masks the whole their universe is not continuous but made up of juxtaposed fragments in their paintings you find them making as much of a halberd as of a human face, as much of an inert stone as of a body in movement drawing a landscape like a geographical map devoting as much care to the figures on a clock as to the statue of Hope or Faith using the same techniques on the statue as on the clock*'.

5 'O. takes a propelling pencil out of his pocket and writes in the margin: Incurable French stupidity.'

6 'The woman is wearing a dark yellow straw hat with the brim held down on either side by a dark-coloured scarf passing over the crown and knotted under the chin. Unruly strands of grey hair escape the headdress and fall across her forehead. All the lower part of the face and the chin jut out as happens with some monkeys and dogs. Below the limp skirt flapping against the calves scrawny ankles are to be seen with black stockings corkscrewed down over them. The feet are shod in big men's boots without laces. The sleeves of the black camisole, dotted with grey diamond shapes, are rolled up and reveal bony forearms covered in yellowish skin. From the end of one of these, extended horizontally and at right angles to the perpendicular line of the body, dangles a rabbit with pearly-grey fur, held by the ears, at times perfectly motionless, at times agitated by spasmodic movements and impotent jerkings of the loins. Emerging from the gnarled yellow fingers of the other hand, the blade of a knife passingly catches the light. The loins of the girl lying in the hay move in time to the rhythmic back and forth movement of the buttocks of the man whose gleaming member catches the light then disappears up to the balls among the thick black shiny hairs, curly as astrakhan. The couple are standing in an area of semi-darkness out of the cone of light thrown across the entry to the narrow brick-walled passage by a reflector mounted on a metal pole with

intersecting girders.'

7 See 'Fiction as process: the later novels of Claude Simon', in *Claude Simon: New Directions*, 1985, pp. 61–74.

8 'we, tossed to right and left, like a floating cork, directionless, sightless'.

9 'La mort de l'auteur', *Mantéia*, 5, 1968. Reprinted (in translation) as 'The death of the author', in *Image, Music, Text*, 1977, pp. 142–8.

10 A helpful summary of these developments is to be found in C. Belsey, *Critical Practice*, 1980, Chapter 3, pp. 56–84.

11 'to read is to attend to the hidden order of the textual work'. 'L'or du scarabée', in *Pour une théorie du nouveau roman*, 1971, p. 51 (first published in 1968). Other examples of this kind of reading are 'La bataille de la phrase', in the same volume, pp. 118–58; and 'La description créatrice: une course contre le sens', in *Problèmes du nouveau roman*, 1967, pp. 91–111 (first published 1961). Robbe-Grillet's insistence on play goes back as least as far as *La Jalousie*, 1957: see Minuit, pp. 82–3. It becomes progressively more pronounced in his theoretical writings: for example, 'Order and disorder in film and fiction', *Critical Enquiry*, autumn 1977, pp. 1–20.

12 'It pleases *The Battle of Pharsalus* to write, from its epigraph, the very recognisable formula *Achilles immobile at full stride*'; 'the text also takes into account the signifier of the epigraph'; 'the frequent arrangements by which the text engenders itself'. Quotations from 'La bataille de la phrase', pp. 124, 127, 118.

13 'from the mass of possibilities will be drawn by preference those which obey this injunction'. 'La bataille de la phrase', p. 129.

14 'it must be agreed that it is in full knowledge of what he is doing that Poe introduces here a new dimension'. 'L'or du scarabée', p. 44.

15 '[T]his is certainly not to say that many an unperceived connection may not have been at work without the scriptor's knowledge.' 'Naissance d'une fiction', in *Nouveau roman: hier, aujourd'hui*, vol. 2, 1972, p. 392. I am indebted to Ed Smyth for first drawing my attention to the ambivalence in Ricardou's attitude.

16 'Every word summons (or commands) others, not only by dint of the images which it draws to itself like a magnet, but sometimes also merely by its morphology, simple assonances, which like the formal necessities of syntax, rhythm and composition, prove as fruitful as its multiple meanings.'

17 For example in 'Réponses de Claude Simon à quelques questions écrites de Ludovic Janvier', *Entretiens: Claude Simon*, 31, 1972, p. 26; 'La fiction mot à mot', in *Nouveau Roman: hier, aujourd'hui*, vol. 2, 1972, pp. 84, 88.

18 'every time that at each of these crossroad words several perspectives, several "figures" present themselves, [the idea] is always to have in mind, in view of the choice about to be made, the initial figure and the

four or five properties derived from it, and never to lose sight of these, failing which . . . there would be no book, in other words unity, and everything would disperse into simple linear succession'. 'La Fiction mot à mot', vol. 2 1972, p. 88.

19 'Interview with Claude Simon', in *Claude Simon: New Directions*, 1985, pp. 13, 15.
20 'gunner', 'to shoot', 'to withdraw', 'charge/load', 'loader/ammunition server/magazine', 'unload/discharge/experience orgasm'.
21 'an integral part of the magma of earth, foliage, water and sky which surrounds him'.
22 ' "Claude Simon", textuellement', in *Claude Simon: colloque de Cerisy*, 1975, pp. 18–19.
23 *Le Nouveau Roman*, 1973, p. 66.
24 *Le Nouveau Roman*, 1973, p. 67; '*Claude Simon: colloque de Cerisy*, 1975, p. 19; 'Le dispositif osiriaque', in *Nouveaux Problèmes du roman*, 1978, pp. 231–2.
25 'Currents of description in *Les Corps conducteurs*', Chapter 2 in *Claude Simon and the Transgressions of Modern Art*, 1988, pp. 134–74.
26 'Currents of description', p. 141.
27 'Currents of description', p. 163.
28 'Currents of description', p. 167.
29 'Le dispositif osiriaque', in *Nouveaux problèmes du roman*, figure 25, p. 237.
30 For a parallel analysis of formalist criticism but from a Derridean perspective, see D. Carroll, *The Subject in Question*, 1982, pp. 161–200. My objection is that it fails to acknowledge the substitution of one kind of hierarchy and coherence for another; Carroll objects that it finds them at all.
31 For example, the interplay of medical motifs from different frames which come together in the fantastic description of an operation (T, 74), 'Currents of description', p. 157.
32 '*emerging* from the other hand . . . one can catch glimpses of the *glint of a knife blade* . . . the man whose *gleaming member glints then disappears* up to the balls . . . out of the *cone of light thrown across the narrow passage* . . . *a reflector mounted on a metal pole*'.
33 See *The Role of the Reader*, 1981, pp. 9–10.
34 'Ainsi ne peut-il [mon trajet] avoir d'autre terme que l'épuisement du voyageur explorant ce paysage inépuisable': Claude Simon, in the preface to *Orion aveugle*, 1970.
35 'Interview with Claude Simon', in *Claude Simon: New Directions*, 1985 p. 13.
36 See 'Authorial correction and *bricolage* in the work of Claude Simon', in *Claude Simon: New Directions*, 1985, pp. 30–49.

37 *After the New Criticism*, 1980, pp. 36 and 58.
38 *Image, Music, Text*, 1977, p. 146.

Chapter 4

1 J. H. Duffy, '*Les Géorgiques* by Claude Simon: a work of synthesis and renewal', *Australian Journal of French Studies*, 21, May–August 1984, p. 161.
2 Quoted from the title of Jean Duffy's article, cited above.
3 '*Les Géorgiques*: "une reconversion totale"? ', *Romance Studies*, 2, 1983, p. 80.
4 The tone of praise for a work of high ambition successfully achieved was set by L. Dällenbach's article, '*Les Géorgiques* ou la totalisation accomplie', *Critique*, 414, 1981, pp. 1226–42. Among the few analyses of *Les Géorgiques* which come close to ignoring its self-reflexive aspects one might cite J. van Apeldoorn's study of the role of animals as a means of initiation to the cyclical pattern of history. 'Claude Simon: mots, animaux et la face cachée des choses', *Cahiers de recherches interuniversitaires et néerlandaises*, 19, 1988, pp. 72–82. The question of genre has been tackled by M. Orr in 'Intertextual bridging. Across the genre divide in Claude Simon's *Les Géorgiques*', *Forum for Modern Language Studies*, 26, 1990, pp. 231–9. M. M. Brewer and A. C. Pugh have considered *Les Géorgiques*'s perceived critique of the writing of history in, respectively, 'Narrative fission: event, history, and writing in *Les Géorgiques*', *Michigan Romance Studies*, 6, 1986, pp. 27–39, and 'Facing the matter of history: *Les Géorgiques*', in *Claude Simon: New Directions*, 1985, pp. 113–30.
5 See 'Bakhtine, le mot, le dialogue et le roman', *Critique*, 239, 1967, pp. 438–65. Reprinted as 'Le mot , le dialogue et le roman', in *Semeiotiké: recherches pour une sémanalyse*, 1969, pp. 143–73.
6 'Bakhtine, le mot, le dialogue et le roman', *Critique*, 239, 1967, p. 441. *Semeiotiké, recherches pour une sémanalyse*, 1969, p. 146.
7 J.-F. Lyotard, *Discours, Figures*, 1978, p. 126.
8 Thus, for example, C. Britton, in discussing Simon's novels: 'Both [texts and pictures] function as anti-representational strategies. This is clearer in the case of intertextual relations which cut across and automatically undermine the illusion of reality by situating the text in relation to other texts rather than to a diegetic referent.' *Claude Simon: Writing the Visible*, 1987, p. 93.
9 *Palimpsestes*, 1981, p. 7.
10 'the work of transformation and assimilation of several texts performed by a centring text which keeps the *leadership* of meaning'. L.

Jenny, 'La stratégie de la forme', *Poétique*, 27, 1976, p. 262.

11 See 'Orion voyeur: l'écriture intertextuelle de Claude Simon', *Modern Language Notes*, 103, 1988, pp. 711–35; or Riffaterre's analysis of a prose poem by Breton which begins: 'An intertext is one or more texts which the reader must know in order to understand a work of literature in terms of its overall significance (as opposed to the discrete meanings of its successive words, phrases and sentences).' 'Compulsory reader response : the intertextual drive', in *Intertextuality: Theories and Practices*, 1990, pp. 56–76 (p. 56).

12 H. Bloom, *The Anxiety of Influence*, 1973.

13 *Communication*, 8, 1966. Reprinted in *Poétique du récit*, 1977, pp. 7–57, and translated in *Image, Music, Text*, 1977, pp. 79–124.

14 'this I which approaches the text is itself already a plurality of other texts, of infinite codes, or more precisely lost codes (whose origins are lost)'. *S/Z*, 1970, p. 16.

15 *Système de la mode*, 1967.

16 The essential, most comprehensive guide to intertextual references in Simon and to intertextual approaches to his work is Mary Orr's *Claude Simon: the intertextual dimension*, 1993. In practice, Mary Orr's own approach remains broadly humanistic and her chapters vary less radically from one another than the sketches of varying methods set out at the beginning of each chapter might lead one to expect.

17 For example, D. R. Ellison: 'The major interpretative difficulty to be faced by the uninitiated non-specialist when reading *Les Géorgiques* is clear enough: how is it possible to understand a discrete work of literature that is composed of references to its own "past" in other texts of the same author and in texts of other writers, to such a degree that the boundaries between it and its antecedents are virtually abolished?', 'Narrative levelling and performative pathos in Claude Simon's *Les Géorgiques*', *French Forum*, 12, 1987, p. 304. Or J. H. Duffy: 'In *Les Géorgiques* intertextual relationships are promoted at the expense of "originality". Individual new developments in *Les Géorgiques* are few and far between; its "novelty" is the thorough stock-taking to which Simon subjects his previous work and his own recognition that its aesthetic success depends more than ever before upon the reader's familiarity with and acceptance of Simon's own priorities and methods', '*Les Géorgiques* by Claude Simon: a work of synthesis and renewal', *Australian Journal of French Studies*, 21, 1984, p. 177.

18 *S/Z*, 1970.

19 'a code of writing accepted in advance by each of the parties, the drawer and the spectator'.

20 'it seems that the artist, making a personal selection of values, has sought, in the proposed scene, to differentiate clearly between the various elements according to their growing importance in his mind';

'this . . . seems to confirm that we have not to do here with an uncompleted canvas, but with a work considered by its author to be quite finished'.

21 In 'From drawing, to painting, to text: Claude Simon's allegory of representation and reading in the prologue to *The Georgics*', *The Review of Contemporary Fiction*, 5 (1), 1985, Anthony Pugh develops with more theoretical subtlety a view similar to the one taken here: that Simon's critique of representation nevertheless engages the reader to enter his fictional world.

22 'He is fifty. He is the general commanding the artillery in the Army of Italy. He is resident in Milan. He wears a tunic trimmed with gilt on collar and breast. He is sixty. He is overseeing the finishing stages of work on the terrace of his country house. He is wrapped up against the cold in an old military greatcoat. He sees black dots. In the evening he will be dead. He is thirty. He is a captain. He goes to the opera.'

23 'In the account which he gives of these events O. tells that at the first shouts a stranger took him by the arm and dragged him away running'; 'he (the cavalryman) reports in a novel the circumstances and the way in which events unfolded in the meantime . . . this narrative can be considered as faithful an account of the facts as possible'.

24 'disagregation or if you prefer complete disintegration of a detachment of officered troops in some hours of a night march'.

25 D. R. Ellison, 'Narrative levelling and performative pathos in Claude Simon's *Les Géorgiques*', *French Forum*, 12, 1987, p. 308.

26 'that eternal re-beginning, that indefatigable patience or doubtless passion which makes it possible to come back periodically to the same places to carry out the same tasks: the same meadows, the same fields, the same vineyards, the same hedges to restock, the same fences to check, the same towns to besiege, the same rivers to cross or defend, the same trenches periodically opened up under the same ramparts'.

27 C. Reitsma-La Brujeere goes so far as to conclude that: 'Ce qui dans *La Route des Flandres* est un échec, est accompli dans *Les Géorgiques*: une reconstruction du passé par une représentation scripturale.', 'Récit et méta-récit, texte et intertexte, dans *Les Géorgiques* de Claude Simon', *French Forum*, 9, 1984, p. 232.

28 This aspect of *Les Géorgiques* has been studied more than any other. See J. van Apeldoorn, 'Une pratique de l'intertextualité: Claude Simon, lecteur d'Orwell', Chapter 3 in *Pratiques de la description*, 1982, pp. 65–145; C. Britton, 'Diversity of Discourse in Claude Simon's *Les Géorgiques*', *French Studies*, 38, 1984, pp. 423–42; J. Fletcher, 'Intertextuality and interfictionality: *Les Géorgiques* and *Homage to Catalonia*', in *Claude Simon: New Directions*, 1985, pp. 100–12; C. Reitsma-La Brujeere, 'Le livre d'Orwell', Chapter 3 in her *Passé et présent dans 'Les Géorgiques' de Claude Simon*, 1992, pp. 63–106; M.

Orr, 'Simon and Orwell' in *Claude Simon: the Intertextual Dimension*, 1993, pp. 161–73.

29 These quotations come from *Claude Simon: Writing the Visible*, 1987, pp. 16, 17. Britton introduces these remarks by saying that 'O's narrative . . . constructs 'O' as a subject who *knows*'. I would say it is the narrator, not O., who knows; and I think this is the view which Britton developed at greater length in 'Diversity of discourse in Claude Simon's *Les Géorgiques*', *French Studies*, 38, 1984, pp. 423–42 (pp. 434–6).

30 'a series of reflex episodes following on one from another'.

31 *Claude Simon: the Intertextual Dimension*, 1993, p. 30.

32 'The Orpheus myth in *Les Géorgiques*', in *Claude Simon: New Directions*, 1985, p. 96.

33 'The Orpheus Myth', p. 98.

34 'not dressed in his general's uniform with its heavy gold braid but his shoulders draped in a toga in classical mode . . . that effigy of a tribune'.

35 'it was like a sort of confused swarming, the confused awakening, the confused commotion of a barbarian horde'.

36 'the living delegation of original, unchanged humanity, specimens unaltered and unalterable, resistant to the centuries, to progress'.

37 'the tranquil determination drawn from reading these Latin authors'.

38 'the accumulation of papers, of old letters and records'.

39 'an apocalyptic dimension on everything which could constitute some cause or pretext for concern'.

40 'an impassive man of stone', thus recalling Pierre in *L'Herbe*.

41 'and it seemed to the boy that he could see it'.

42 'the powerful, muscular body beginning to take on weight, naked, like these drawings copied from classical models'.

43 'To be descended from Faulkner? That doesn't embarrass me. We all descend from someone.' M. Chapsal, 'Entretien. Claude Simon parle', *L'Express*, 564, 13–19 April 1962, pp. 32–3.

44 *Tel Quel*, 2, 1960, p. 95.

45 'under the melancholy and silent murmur of an autumn rain'.

46 'Because he had a brother . . . ', and the boy: 'A br. . . What brother?', and Uncle Charles: 'Legally and biologically. Yes. Because it is conventional to give that name to the product of two embryos issued from the same male glands and grown in the same womb. Except that they resembled one another in just about the way a photographic negative resembles its print. In other words exactly the same and exact opposites . . .', and the boy: 'So he had a brother? But why . . .', and Uncle Charles: 'You mean: why has he never been mentioned? Well, that's it: precisely!'

47 *Absalom, Absalom!*, Chatto & Windus, 1965, pp. 172.

48 It would be desirable but difficult to prove such assertions empirically.

My own testing has been on a sample of mature postgraduate students, too small to be conclusive. Do hostile reviews prove the contrary? Too many Parisian reviewers continue to read Simon as if they thought he ought to conform to their hazy recollections of Robbe-Grillet's theories of the 1950s. See A. Rinaldi, 'Claude Simon: un catalogue cousu de fil blanc', *L'Express*, 12–18 September 1981, pp. 72–3, and B. Poirot-Delpech's curiously ambivalent review of *L'Acacia*, 'Entre deux guerres', *Le Monde*, 1 September 1989. And abroad? I can at least report that in November 1992, according to the book programme of Südwestfunk, Stuttgart, the newly translated *Georgics* topped the list of foreign novels recently reviewed and recommended in the German press ('Claude Simon an der Spitze', *Bietigheimer Zeitung*, 27 November 1992).

49 Both Mary Orr in *Claude Simon: the Intertextual Dimension*, 1993, p. 3, and Cora Reitsma-La Brujeere in *Passé et présent dans 'Les Géorgiques' de Claude Simon*, 1992, pp. 7–8, use signposting as their main criterion for determining the presence of an intertext in Simon's fiction.

50 See M. Orr, 'Lytton Strachey: literary embellishment and functional intertext in Claude Simon's *Les Géorgiques*?', *French Studies Bulletin*, 26, spring 1988, pp. 14–17; and G. Neumann, 'Claude Simon et Michelet: exemple d'intertextualité génératrice dans *Les Géorgiques*', *Australian Journal of French Studies*, 24, 1987, pp. 83–99.

51 C. Reitsma-La Brujeere, *Passé et présent dans 'Les Géorgiques' de Claude Simon*, 1992, p. 30.

52 This relationship is related to the form of intertextuality which Genette in *Palimpsestes* calls metatextuality, in which the host text comments on the intertext. In this case however there are no signposts to indicate that *S/Z* is present in *Les Géorgiques*. The metatextual relationship is between these two texts and mine.

53 'the missives sealed with wax which she opened, deciphered or rather decoded, trying to see in what he called the blue sector,the green sector, the pink sector, those poplars, acacias, fields, vineyards sent as it were by post on these rectangles of paper covered with little signs from which (after the fashion of those microscopic Japanese flowers which, when thrown into water, swell, unfold into unsuspected crowns of petals) would materialise once more the demanding earth, the slopes, the valleys turn by turn verdant, russet, parched or muddy under the changing skies, the slow drift of the clouds, the dew, the storms, the frosts, in the unchanging alternation of the unchanging seasons'.

54 'Le titre surdéterminé des *Géorgiques* renvoie donc aussi à la geste de Georges', '*Les Géorgiques* ou la totalisation accomplie', *Critique*, 414, 1981, p. 1241.

Chapter 5

1 Introductory note to Book One of *Les Confessions*.
2 'L'essence et les sens', in *Pour une théorie du nouveau roman*, 1971, pp. 200–11.
3 *Moi aussi*, 1986, pp. 30–1.
4 *Lire Leiris. Autobiographie et langage*, 1975; 'L'ordre du récit dans *Les Mots* de Sartre', in *Le Pacte autobiographique*, 1975, pp. 197–243; *La Mémoire et l'oblique: Georges Perec autobiographe*, 1991.
5 in 'Fragment autobiographique imaginaire', *Minuit*, 31, 1978, p. 2.
6 'To 'express onself' is an obscene term. To use it in public is to run the risk of being hacked to pieces. Use it in the presence of Robbe-Grillet or Ricardou and you sign your death warrant. The author is dead, may the God of criticism receive his defunct soul. It's language which speaks aloud, alone, no one speaks, it says nothing, without source or origin, it unfurls in great symbolic forms. *Le Livre brisé*, 1989, pp. 72–3.
7 'Autobiography as intertext: Barthes, Sarraute, Robbe Grillet', in *Intertextuality: theories and practices*, 1990, pp. 108–29 (p. 116).
8 *Simon: 'Histoire'*, 1982, p. 11.
9 See A. C. Pugh, 'Claude Simon et la route de la référence', *La Revue des sciences humaines*, 84 (220), 1990, pp. 31, 42.
10 'from *L'Herbe* on, all my novels are virtually autobiographical'. Quoted in English translation in A. Duncan, 'Interview with Claude Simon', in *Claude Simon: New Directions*, 1985, p. 16.
11 'it's not me, I mean a faithful or exhaustive portrait of me; that's not possible'. D. Eribon, 'Entretien. Fragments de Claude Simon', *Libération*, 29–30 August 1981, p. 21.
12 'I am now an old man and, in common with many other inhabitants of our old continent of Europe, the first part of my life was fairly full of incident: I have witnessed a revolution, I have been to war in particularly murderous circumstances (I belonged to one of those regiments that General Staffs coolly sacrifice in advance and after a week there was almost nothing left of it), I have been taken prisoner, I have known hunger, physical labour to the point of exhaustion, I have escaped, I have been critically ill, several times on the point of death from violent or natural causes, I have rubbed shoulders with the most varied of people, from priests to men who burnt down churches, from peaceable bourgeois to anarchists, from philosophers to people who couldn't read or write, I have broken bread with men on the wrong side of the law, I have travelled all round the world . . . but never yet, at the age of seventy-two, have I discovered any sense to all this, beyond what, I think, Barthes said, following Shakespeare, that "if the world signifies anything, it signifies nothing" – except that it is.'
13 'I am a man who is trying to live, I am entirely absorbed in this difficulty

of living, I am searching for what can help me to go on and for that I have to find something solid on which I can count.'

14 'the three of four basic needs, like sleeping with women, eating, talking, procreating, for which men are made and which they can't do without'.

15 'To say that the world is absurd amounts to confessing that one persists in believing that there is a reason.'

16 'the world is neither significant nor absurd: it quite simply *is*'. 'Une voie pour le roman futur', *Nouvelle Revue Française*, July 1956, pp. 77–84. Reprinted in *Pour un nouveau roman*, 1963, p. 21.

17 'I don't explain, I note, and I restrict myself to telling what I have seen.'

18 'an ideal vision of the world, voluntarily limited and arbitrary, each of these visions being based on a conception, or rather an ethic of the visible universe, or even, much more serious for painters, simply a code of morals'.

19 'a universe for the first time devoid of signposts. So completely stripped of everything except truth and coherence that for the first time we were offered the visible world, in its total magnificence, without commentary or limitation, and through it, the world simply as it is'.

20 'In the past I used to lie late in bed and I felt good. I would smoke cigarettes, enjoying my outstretched body. . . . In Paris, framed by the window, there was the side of a house, a dome . . . looking at it I could travel and remember mornings waking up in hotel bedrooms of unfamiliar towns.'

21 'The sounds and colours intermingle . . . everything becomes tangled and runs into everything else . . . Because of all this, I am not myself.' 'You might as well try to stop water running between your fingers. Try. Try to find yourself. "I is another." Not true: "I is others." Other things, other smells, other sounds, other people, other places, other times.'

22 'My subject won't wait. What is your subject then? A sprint. In what sense? People and a heap of things, smells, times, ideas, running figures, and me in the middle of them, running out of breath to keep up with them. You mean that if you stop, you'll have forgotten what you wanted to talk about? You never know what you're talking about until you start doing it.'

23 See *Pour un nouveau roman*, 1963, pp. 26, 82.

24 For some readers of the time, the psycho-pathological excluded freedom from Robbe-Grillet's early works, for example J. Alter, *La Vision du monde d'Alain Robbe-Grillet*, 1966; others found his novels to be controlled studies of such states, notably B. Morrissette in *Les Romans de Robbe-Grillet*, 1963 and 1971.

25 'this kind of story, without start or finish, gives readers the pip'.

26 'one of these hairy trunks still to be found in attics'.

27 'she was hoping to find neither diary, nor memoirs, nor yellowed

letters, nor anything of that sort . . . because those were kinds of ideas (keeping a diary, writing the story of your life) which couldn't possibly have even crossed the mind of the woman who had kept them'.

28 'And, a little later, when she had opened the box, standing there looking at its contents, without touching them, still feeling the same perplexity, the same alarm, frowning, silent, staying like that for a quarter of an hour perhaps, motionless.'

29 'Le général Willot fut battu, et sa gauche eût péri toute entière, si l'énergie d'un bataillon du régiment de Champagne, commandé par le lieutenant-colonel Sauret, n'avait donné aux fuyards le temps de gagner le Boulon (20 avril 1793). Les représentants de peuple ordonnèrent à Willot d'aller à Toulouse rendre compte de sa conduite et suspendirent la Houlière à cause de son grand âge. Cet infortuné vieillard se brûla la cervelle.' 'General Willot was defeated, and his left wing would have perished utterly, had not the energy of a battalion of the Champagne Regiment, commanded by Lieutenant-Colonel Sauret, given the fugitives time to reach Le Boulon (20 April 1793). The Representative of the People ordered Willot to go to Toulouse to give an account of his conduct and suspended la Houlière because of his advanced age. That unfortunate old man blew his brains out.' P. Vidal, *Histoire de la révolution française dans les Pyrénées orientales*, vol. 2, 1886, p. 137.

30 D. Bourdet, 'Images de Paris: Claude Simon', *La Revue de Paris*, January 1961, pp. 136–41. Reprinted in *Brèves Rencontres*, 1963, pp. 215–24 (p. 218).

31 'About two hours later he is in the saddle again next another trooper behind the leader of the squadron and a lieutenant (in a novel he reports the circumstances and the way things happened in the interval: making allowances for the weakened state of his perceptive faculties due to weariness, lack of sleep, noise and danger, the inevitable gaps and distortions of memory, that narrative can be considered as faithful an asccount of the facts as possible: the crossroads and the fields strewn with corpses, the wounded man covered in blood, the dead man stretched out on the far side of the ditch, his own gradual return to consciousness, sudden decision, panting sprint up the hill through the meadows separated by hawthorn hedges, crossing of the road patrolled by enemy armoured vehicles, march through in the forest (*And where will you go?*), thirst, the silence of the undergrowth, the singing of the cuckoo, the distant sound of bombardments, the unexpected meeting with the two officers who had escaped from the ambush, the order casually given him to mount one of the two horses led by the orderly, the crossing of the shelled town, etc.'

32 '*so for a long time the idea of death will remain associated in his mind with the fragrance of the sponges soaked in eau de Cologne which were put into their hands when as children, during Holy Week, dressed as*

penitents, they were led before the emaciated Christ-figures displayed on beds of flowers whose toes they wiped before honouring them with a kiss. . . . Later the idea of death will become confused for him with the sickening smell of hot rancid oil which permeated the food served to the foreign volunteers in the great dining room of the requisitioned luxury hotel in Barcelona. . . . Later still and for many years, this same idea of death will for him become inseparable from the names of a series of hamlets or villages between the Meuse and the Sambre, abandoned in the abandoned countryside, shaken now and then by the echoes of the explosions.'

33 *Les Mots*, 1964, p. 165.

34 'Suddenly, while he is writing, he realises that he is describing not what he experienced, but, he says, an engraving representing that event, an engraving which he saw five or six years later and which (these are the terms he uses) has since "taken the place of the reality".' A. Poirson and J.-P. Goux, 'Un homme traversé par le travail. Entretien avec Claude Simon', *La Nouvelle Critique*, 105, June–July 1977, p. 37.

35 'Claude Simon: fiction and the question of autobiography', *Romance Studies*, 8, Summer 1986, p. 84.

Chapter 6

1 'Ce qu'il nous importait avant tout de marquer en montrant que le texte simonien "implique" la psychanalyse, c'est que cette fiction des origines ne peut qu'impliquer le lecteur lui aussi: sauf à se donner le change, est-il vraiment assuré de savoir qui il est?' 'What we most wanted to draw attention to in showing that Simon's texts 'imply' psychoanalysis, is that this fiction of origins cannot help but 'implicate' the reader: unless we fool ourselves, can we be really sure of who we are?' *Claude Simon*, 1988, p. 94. Dällenbach is playing on the double meaning of the verb *impliquer*: to imply and to implicate.

2 Psychoanalytic theory informed Stephen Heath's study of Simon's realisation of identity in language in *The Nouveau Roman. A Study in the Practice of Writing*, 1972, pp. 153–78; and the contributions of Sylvère Lotringer and Irène Tschinka to the Simon colloquium at Cerisy in 1974. Anthony Pugh's first major article in this area was 'Du *Tricheur* à *Tripyque*, et inversement', *Etudes littéraires*, 9, 1, avril, 1976, 137–60. Books of the 1980s which have given a substantial place to psychoanalytic criticism include Pugh's *Simon: 'Histoire'* and David Carroll's *The Subject in Question: the Languages of Theory and the Strategies of Fiction*, both 1982; Britton's *Claude Simon: Writing the Visible*, 1987, and Dällenbach's *Claude Simon*, 1988.

3 *Simon: 'Histoire'*, 1982, p. 19. See also p. 12: 'What we probably mean, therefore, when we describe a work as "authentic", is that it has

produced in us echoes of our own experience.'

4 Dällenbach speaks not only of the reader's involvement in the search for origins (*Claude Simon*, 1988, p. 94) but also of 'le roman familial qui est assurément le nôtre', p. 68. On pp. 22 and 23 of *Claude Simon: Writing the Visible*, 1987, Britton elaborates and applies to Simon Freud's theory of the pleasure of the text as set out in his essay 'Creative writers and day-dreaming', *Collected Works*, vol. 9, pp. 143–153.

5 Freud, especially, *On Sexuality*, vol. 7, 1977; J. Laplanche and J. B. Pontalis, *Vocabulaire de la psychanalyse*, 1973; M. Robert, *Roman des origines et origines du roman*, 1972; J. Lacan, *Ecrits*, 1966, and *Ecrits: a selection*, trans. A. Sheridan, 1977; C. Britton, *Claude Simon: Writing the Visible*, 1987.

6 See Lacan, 'The agency of the letter in the unconscious or reason since Freud', in *Ecrits: a selection*, 1977, pp. 146-78; and also Chapter 3, 'Language and the Unconscious', in M. Bowie, *Lacan*, 1991, a lucid exposition of Lacan to which I am indebted throughout this chapter.

7 For example in *The Subject in Question: the Languages of Theory and the Strategies of Fiction*, 1982, Chapter 3, 'Psychoanalysis and Fiction or the Conflict of Generation(s)', David Carroll uses Derrida and an analysis of *Le Sacre du printemps* to argue that 'the Imaginary underlies the Symbolic as much as the Symbolic the Imaginary', p. 49.

8 For a sample of Klein's views, see 'The Early Stages of the Oedipus Conflict' and 'The Importance of Symbol Formation in the Development of the Ego', in J. Mitchell (ed.), *The Selected Melanie Klein*, pp. 69–83, 95–114.

9 Pugh, *Simon: 'Histoire'*, 1982, p. 49.

10 Dällenbach, *Claude Simon*, 1988, p. 91.

11 Dällenbach, *Claude Simon*, 1988, p. 62.

12 See Freud, 'Family romances', in *On Sexuality*, p. 223; and Robert, *Roman des origines et origines du roman, 1972*, pp. 50–1.

13 For example, J. Duffy, '(Ms)reading Claude Simon: a partial analysis', *Forum for Modern Language Studies*, 23, 1987, pp. 228–40; W. Woodhull, 'Reading Claude Simon: gender, ideology, representation', *L'Esprit créateur*, 27, 1987, pp. 5–16.

14 'It seems to me that the representation of the female characters is so transparently determined by a configuration of half-acknowledged fears and desires that the question of attaching any kind of objectivity to them simply does not arise: they are an extreme case of the visible as object of phantasy, and hence of fictional representation as phantasy.' Britton, *Claude Simon: Writing the Visible*, 1987, p. 167.

15 '[A]s if, sitting there at that moment of abolished time beside Rose's dead body, enclosed, buried in that flesh, that heavy fragrance of lilac slowly wilting, withering, he found himself returned to a state which was in some sense foetal, curled in the (supposedly) aching, tormenting

quietude of a life within the womb from which – for the second time, and for the second time from between the thighs of a woman, although this one was five years younger than he – he was going to be expelled, projected, crying and terrified, into the void.'

16 Britton, *Claude Simon: Writing the Visible*, 1987, p. 77.

17 Cf. Anthony C. Pugh, who argues that in *Le Tricheur* the signifiers of father and mother are interchangeable and that, more generally, 'la mère . . . nous le verrons, assume à la place des pères défunts et déchus, le rôle de détenteur de la Loi du récit'. 'Du *Tricheur* à *Triptyque*, et inversement', *Etudes littéraires*, 9 (1), avril 1976, p. 144.

18 'a moving network of specks of light and shade ceaselessly forming and dissolving itself'.

19 Britton, *Claude Simon: Writing the Visible*, 1987, p. 125.

20 Pugh, *Simon: 'Histoire'*, 1982, pp. 52, 53.

21 'the mother's beliefs and the catechism classes'.

22 Lambert is a Swann to the narrator's Marcel: his early linguistic revolt peters out in political clichés, first of the left, then of the right.

23 It is highly debatable and much debated by psychoanalysts as to whether any such phase exists. See J. Laplanche and J. B. Pontalis, *Vocabulaire de la psychanalyse*, pp. 323–5. The nostalgia for it certainly exists as a literary phenomenon, as Marthe Robert demonstrates in the case of *Don Quixote* (*Roman des origines et origines du roman*, 1972, pp. 105–234).

24 'my double still unsteady on emerging from the maternal dark, fragile, soiled, protesting and miserable'.

25 'civilisation is cursed as the work of the creator father who set history in motion; it belongs to the category of evil and separation because, having imposed on the world the double curse of sexuality and appropriation of the land, it breaks for ever the union of the little child with "the sacred womb of our first mother" '. Robert, *Roman des origines et origines du roman*, 1972, p. 203.

26 'we can't lose or destroy each other'.

27 'thinking that I could go on like this, advancing still, disappearing into the warm womb of night, seeing nothing other than the reassuring darkness'.

28 'the woman bending forward the mysterious upper part of her body its white flesh shrouded in lace the womb which perhaps was already carrying me in its dark tabernacle a sort of gelatinous tadpole coiled on itself with its two enormous eyes silkworm's head toothless mouth gristly insect's forehead, me?'

29 Pugh, *Simon: 'Histoire'*, 1982, p.45.

30 In *Claude Simon: Writing the Visible*, 1987, mainly Chapter 2, pp. 18–43.

31 Britton, *Claude Simon: Writing the Visible*, 1987, p. 34.

32 Britton, *Claude Simon: Writing the Visible*, 1987, p. 65.

33 'The gunner can still easily read the bold print of the paragraph heading but he has to take the book over to the window to read the fine print of the text: Thus far we have spoken only of the action of the continental waters on stones and land, but sea waters may also work their effects. The wind blowing over the surface of the sea produces waves, which sometimes hurl themselves violently against the coasts. The sea thus makes inroads on the edges of the land, causes rocks and earth to cave in, rips away the hardest stones and rolls them in its waters. As is the case with the contintental waters, rocks are destroyed more or less quickly, depending on the degree of their resistance. When the same rock has some dense and some softer parts, the former, less quickly demolished by the waves form columns or pillars in the sea. Thus at Étretat (fig. 111), Dieppe, etc., the chalk which forms cliffs at the sea edge has some parts which crumble easily and others which are more solid. The former disintegrate more quickly and that is why one can see forming at the coast great arcades of rock supported by the harder parts which constitute the pillars. In the foreground of the picture, the lines which represent the sea are further apart and at the same time thicker, sometimes leaving elongated, slightly swollen, white spaces, like lightly rolling billows of foam. Within each wave, the shingle forming the sea bed seems to rise in the transparent water, like a carpet being rolled, then the crest of the wave breaks and it reappears in its original place, to rise again. The little waves expire one by one on the foreshore with a cool sound. It seems that here and there gleams of pink play on the pale green water. The ammunition server jogs the gunner's elbow and says Hey Charlie I'm talkin' to you d'you hear bloody hell fine time to have your head in a book what you readin'? He wrenches the book from his hands, turns the pages towards the window and reads the bold print of the heading: 145. DESTRUCTION OF COASTS BY WAVES. He says bloody hell and what about the destruction of cunts like us where d'they talk about that? Angrily he throws down the book. While one of the masons holds the ends of the planks chest-high, the other, squatting below, struggles to pull towards him one of the heavy trestles whose feet are occasionally blocked by a heap of rubble, a wedged brick or stone which he has to remove by hand. Finally, he gets to his feet again, gestures to the other workman that he can let the planks fall and leaves the room to return shortly afterwards pushing an iron wheelbarrow with an inflatable tyre. Cross-wise on the wheelbarrow lies a building labourer's spade with rounded, chipped blade. She feels his moustache on her neck below her ear she feels his lips she suddenly feels his moist rasping tongue on her skin she shivers, with her hand flat against his chest she pushes him away, she says no leave me alone I forbid you no, she straightens her back against the gate and pulls her face away. In the dark it appears vaguely oval, bluish, its features signalled only by two dark patches marking the hollows of the eyes and a larger patch at the mouth. He throws down his cigar which lands on a stone in the road and a brief shower of sparks scatters in the darkness.'

34 'Les aventures du lecteur', *Poétique*, 29, 1977, p. 83.

35 Britton, *Claude Simon: Writing the Visible*, 1987, pp. 66–7.

36 'The description (the composition) can be continued (or completed) almost indefinitely depending on the attention to detail with which it is executed, the force of the proposed metaphors, the addition of other objects visible in their entirety or fragmented by attrition, or time, or shock (or else because they appear only in part in the frame of the picture), not counting the various hypotheses to which the spectacle can lead. Thus it has not been said whether (perhaps through a door opening to a corridor or to another room), a second, stronger bulb is not lighting the scene, which would explain the presence of opaque (almost black) shadows which extend over the tiled flooring from the visible (described) or invisible objects – and perhaps also the stretched and stilt-like shadow of a person standing in the frame of the door. Nor has any mention been made of the sounds or the silence, or of the smells (gunpowder, blood, dead rat, or simply the subtle, dying, rancid scent of dust) which dominate or are perceptible in the place, etc., etc.'

37 M. Evans, *Claude Simon and the Transgressions of Modern Art*, 1988, p. 231.

38 'He clasps and unclasps his hand or passes his palm lightly over the elastic nipple. . . . Without ceasing to caress her, he gazes at the fall of milky flesh.'

39 'that breast which I glimpsed for the first time, discovering that beneath the mysterious black silks were to be found skin, flesh whose softness, whose extreme whiteness heightened still more their unreality, their sacred character as of things destined to remain hidden and on which – as on the Host at the moment it is raised – one must not look'.

40 Britton, *Claude Simon: Writing the Visible*, 1987, pp. 123–4.

41 Lacan discusses Freud's remark in *Le Séminaire, livre 11: les quatre concepts fondamentaux de la psychanalyse*, 1973, p. 45; also in *Ecrits: a selection*, 1977, p. 44.

Chapter 7

1 'To decipher *L'Acacia* is a most pleasurable experience.' 'Un drôle d'arbre: *L'Acacia* de Claude Simon', *Romanic Review*, 82, 1991, p. 220. Sarkonak also summed up another way: 'l'oeuvre de Simon – une encyclopédie sauvage, son index – *L'Acacia*', p. 212. To which one must add: 'its indexer – Sarkonak'.

2 **'– So that was the colonel you described in *La Route des Flandres*. He**

existed then?

– Yes, and I refer you to my account: the confusion , those in front surging back, those behind trying to get past, the stupid order given by our lieutenant: 'Dismount. Prepare to fight on foot.' (A manoeuvre impossible to execute under fire and which would allow the enemy to carry out a massacre: so you can see that the complete slackness of command held sway at every level of the hierarchy, from generals to colonels, and including lieutenants – one must however be fair to our general; he had the grace to honourably blow out his brains . . .), then, immediately afterwards, the countermanding order: 'Mount and gallop!', and then my girth-strap too long for the little mare that I'd been given to replace the one I'd done in to get back across the Meuse, my saddle turning as I put my foot in the stirrup, my sprint holding the mare by the bridle, the shock (a horse, the blast of a shell?) which made me lose consciousness, and when I came round, I found myself on all fours in the middle of a road, surrounded by dead or wounded men and horses. My race once more for the nearest hedge, how I managed to get past between the patrolling armoured vehicles, my march westwards through the forest, towards the line of defences which I knew was there and where I imagined I'd be received with congratulations, and when at last I got there, amazement: silence, nothing but the distant sound of guns or explosions, dazzling springtime nature, birdsong and – *NOBODY*. I can bear witness to that: on the morning of 17 May 1940, there was *NOBODY* in the defence works north of Solre-le-Château! The defences had been purely and simply abandoned without a fight, intact, without the slightest trace of a shell or any bombardment whatever.' 'La déroute des Flandres', *Le Figaro*, 13 July 1990, p. 11.

3 'Just as he couldn't say either how long he remained unconscious on what couldn't exactly be called a battlefield (the crossing of two local roads amid ripening corn and meadows in flower): all he can remember (or rather can't remember – it will only be later, when he has time: for the moment he is solely occupied in careful observation of the countryside about him, in reckoning the distance which separates him from the next hedge, while he lifts the strap of his carbine over his head, opens the breech, tips it up and withdraws it) are the still faint and transparent shadows of horses on the ground, slightly in front and to the right, so elongated by the first rays of the sun that they seem to move without advancing, as if mounted on stilts, raising their elongated, grasshopper's legs and lowering them again so to speak at the same place, like a fantastic animal seeming to imitate on the spot the action of walking, the long column of cavalrymen retreating, dozing still as they emerge from night, backs bent, bodies swaying backwards and forwards in the saddle, the head of the column turning right at the crossroads, then suddenly the shouts, the bursts of machine-gun fire,

the head of the column surging back, then other machine-guns behind, the rear of the column starting to gallop, the riders mingling, bumping into one another, confusion, uproar, disorder, continued shouting, detonations, conflicting orders, getting ready to remount the mare from which he has jumped down, the foot on the stirrup, the saddle turning, and now braced, pulling and pushing with all his might to put it back in place, fighting against the weight of sabre and saddlebags, the reins pushed to the crook of his left elbow, jostled, gashing his palm on the prong of the buckle, deafened by the explosions, the galloping of horses, or rather perceiving (through hearing, sight) something like fragments which succeed, displace, unmask, clash with one another, whirling round: horse flanks, boots, hooves, rumps, falls, fragments of shouts, of sounds, the air, the space as if fragmented, chopped into tiny pieces, shredded by the crackle of the machine-guns – then giving up, starting to run, still cursing, among the crazed horses, the shouts, the uproar, the mare which he is holding by the bridle moving at a canter, the saddle under her belly, then suddenly nothing more (not even feeling the shock, no pain, not even the awareness of stumbling, falling): darkness, no more sound (or perhaps a deafening din cancelling itself out?), deaf, blind, nothing, until slowly, emerging little by little like bubbles on the surface of murky water, appear vague blurred shapes which run into one another, fade, then reappear again, then become clear: triangles, polygons, stones , tiny blades of grass, the macadamised surface of the road on which he is now supporting himself on hands and knees.'

4 'based on lived experience'. M. Alphant, ' "Et à quoi bon inventer?" ', *Libération*, 31 August 1989.

5 M. Alphant, 'La route du Nobel', *Libération*, 10 December 1985, p. 28. The factual information which follows in this paragraph has been confirmed to me by Claude Simon.

6 'Comme toutes les contraintes, celle de renoncer à la fiction est très fertile.' 'Like all constraints, giving up the fictitious is very productive.' Claude Simon, interviewed by M. Alphant, *Libération*, 31 August 1989.

7 In the *Figaro* interview Simon portrayed Colonel Rey, commanding officer of the 31st Dragoons, in a manner at once so lifelike and unflattering that it provoked a strong letter of protest from the Colonel's daughter. *Le Figaro*, 27 July 1990, p. 5.

8 Claude Simon's parents were married on 8 February 1910. According to Captain Simon's 'Livret matricule d'officier', the Captain was stationed in France from 9 December 1909 until 25 April 1912 when he embarked for Madagascar. Captain Simon returned from Madagascar, disembarking at Marseilles on 4 June 1914. He was given three months' leave ('congé de convalescence'). The 'livret d'officier' states that

Captain Simon was mobilised on 6 or possibly 2 August 1914.

9 A. C. Pugh reproduces Simon's sketch-map and factual account of the incident and comments on its fictional appearances in *La Corde raide*, *La Bataille de Pharsale* and *Leçon de Choses* in 'Claude Simon et la route de la référence', *La Revue des sciences humaines*, 84 (220), 1990, pp. 23–45 (pp. 41–5).

10 'Vertige et parole dans l'oeuvre de Claude Simon', in *Une parole exigeante*, 1964, pp. 89–110. John Sturrock also dealt with history thematically in *The French New Novel*, 1969, pp. 43–103. More recent essays adopting a comparable aproach towards a wider range of novels or specific later novels include P. Daprini, 'Claude Simon, History and "L'Innommable réalité" ', in *Literature and War*, ed. D. Bevan, 1990, pp. 167–78; F. Châtelet, 'Une vision de l'Histoire', *Critique*, 37, November 1981, pp. 1218–26; F. Claudon, '*Histoire* et la représentation de l'histoire chez Claude Simon', *Beiträge zur Romanischen Philologie*, 26, 1987, pp. 253–8. A. C. Pugh has considered both the relationship of Simon's novels to historical reality and their critical contribution to historiography, for example in 'Facing the matter of history: *Les Géorgiques*', in *Claude Simon: New Directions*, 1985, pp. 113–30, and in 'Defeat, May 1940: Claude Simon, Marc Bloch and the writing of disaster', *Forum for Modern Language Studies*, 22, 1985, pp. 59–70. Historiography has also been Jean Duffy's concern in 'The subversion of historical representation in Claude Simon', *French Studies*, 41, 1987, pp. 421–37.

11 'perhaps, after having struck a first blow, shaken with horror, history granted itself a breathing space'; 'history itself seeing to what remained . . . acting (history) with its terrifying excessiveness, its incredible, heavy humour'.

12 *Claude Simon: Writing the Visible*, 1987, pp. 147, 150.

13 I am indebted here to D. Bebbington, *Patterns in History. A Christian Perspective on Historical Thought*, 1990, Chapter 4, 'The idea of progress', pp. 69–91, and Chapter 7, 'The philosophy of historiography', pp. 140–67.

14 'the mass production of refrigators, cars and radios'.

15 K. Marx, *The German Ideology*, in *Karl Marx: Selected Writings*, ed. D. McLellan, 1977, p. 165.

16 Karl Marx, *The Holy Family*, in *Karl Marx: Selected Writings*, 1977, p. 139.

17 'that damned dear old bearded lady who foresaw everything'.

18 'URINE – SPITTLE – BLOOD'.

19 'that nitro-glycerine in the form of printed paper'; 'a world in which violence, rapacity and murder have always been at home'.

20 *Karl Marx: Selected Writings*, 1977, p. 300.

21 'history . . . beginning to run light, racing, veering towards parody, to

the comical . . . the invisible director in haste to finish with it, oppressed by the repetitions of a play performed a hundred times, scarcely leaving the actors time to get their speeches out, already gesturing the next one on, tyrants, despots for a month, a week, a day, dead the evening after'.

22 'the perpetuation, a living delegation of original, unchanged humanity'; 'a sort of dried and wrinkled fruit, forgotten by history'.

23 'the old *landsknechts*, *reiters* and cuirassiers of long ago'.

24 'a sort of secret, demanding, imperious life, like what forced the insects to swirl on the spot, suspended, mingling ceaselessly, aimlessly, their invisible flight-paths, their suspended, immobile, obstinate cloud, each of its particles in perpetual motion, bright, golden against the background of greenery, then (the same cloud, or another) standing out darkly against the fading light of the sky'.

25 Marx writes of man always having before him not 'an antithesis of nature and history' but 'an historical nature and a natural history'. *The German Ideology*, *Karl Marx: Selected Writings*, 1977, p. 174. Le Roy Ladurie has written a two-volume *Histoire du climat depuis l'an mil*, 1990.

26 'He was a fairly big, burly man, with regular features, a moustache curling up at the tips and square-cut beard, and his pale, china-blue eyes, staring out of the peaceful bloodstained face, gazed upwards at the bullet-torn foliage in which the summer afternoon sun was playing. On his tunic the sticky blood made a bright red stain, already brown at the edges where it was beginning to dry and almost hidden from view by the swarm of striped flies with grey, black-speckled wings which were jostling and clambering over one another, like the flies which settle on excrement in undergrowth.'

27 'with his back against the tree like a medieval knight or a colonel of the Empire'; 'the bullet "taken full in the forehead" '.

28 'You would have said that she had no desires, no regrets, no plans. Cheerful on the whole, reported those who knew her at that time, fond of food (and therefore no doubt sensual).'

29 'She was religious. At least every Sunday she accompanied her mother and her sister to the cathedral'; 'She played tennis (at least a photograph showed her on a tennis court).'

30 'Among those who fell in the fighting of 27 August was a forty-year-old captain whose still warm body had to be abandoned at the foot of the tree against which it had been propped.'

31 'the girl', 'the widow'; 'the young scholarship boy', 'the captain'; 'the child', 'the tourist', 'the cubist painter', 'the reservist', 'the corporal'.

32 'Their education, memories of staggering shapes at evening in the village street, cries of women being beaten, swaths of corn from which a slender lash would sometimes fall wriggling, years in mountain schools round which the snow melted only to give way to rain, the loss of a

cousin whose handkerchief they remembered stained with blood, were to leave the two primary schoolteachers until the hour of their death with an almost superstitious feeling somewhere beween fear and a visceral repugnance which made them confuse in the same terror (lowering their voices if by chance they happened to mention them, as if the words themselves were charged with a malign, soiling power, as if obscene) drunkenness, adders, mud, priests and tuberculosis.'

33 When I was bold enough to suggest in print that the great-uncle might have been invented for the sake of symmetry (in 'le nouveau roman jurassien', *Le Jura français*, 58, 1991, p. 10) Simon reacted sharply: 'on était fier dans ma famille paternelle de cet arrière-grand-père (ou oncle – cela je ne sais au juste . . .) qui s'était caché pour échapper aux sergents-recruteurs de Napoléon'. 'my father's family were proud of the great-grandfather (or uncle – that I don't know exactly . . .) who had hidden to escape Napoleon's recruiting sergeants'. Letter of 12 July 1990.

34 'we're all off to the same place: to cop it'.

35 Wilhelm Windelband, quoted in Bebbington, *Patterns in History*, 1990, p. 147. For this account of historicism I am indebted to Bebbington, Chapter 5, 'Historicism', pp. 92–116, and to M. Levinson, 'The New Historicism: Back to the Future', in *Rethinking Historicism: Critical Readings in Romantic History*, 1989, especially pp. 28–31.

36 The general was Rommel, as is clear from the context (L'A, 328–9). Rommel recounts this incident in his memoirs, published in French as *La Guerre sans haine*.

37 'Let each say frankly what he has to say: the truth will arise from the converging sincerity of these words.' *L'Etrange défaite*, 1990, p. 54.

38 'china-blue eyes' and 'square-cut beard'; 'chapped hands' and 'furrowed faces'.

39 'something which was unlike a charge, unlike anything they had ever learned from books or from experience'.

40 'thinking later But for the part which he had destined us to play that had no importance I could just as well have been tied astride a donkey with my head facing its tail'.

41 'the series of bereavements which had struck the family'.

42 'that adored woman buried in nothingness since so long ago and whose memory after twenty years still breaks my heart'.

43 'the titanic exploit of giving birth to a world and killing a king'.

44 ' "Pass it on: the German armoured cars are in the village! Stop! Armoured cars in the village! Pass it on! The Ge . . ." '

45 'There's no firing to be heard any more. Squatting now, he looks around him.'

46 M. Bowie, *Lacan*, 1991, p. 18.

47 T. G. Pavell, *The Feud of Language: a History of Structuralist Thought*, 1989, p. 8. On pp. 1–3 of the same work, Pavell gives a short roll-call of

those who contributed to the debate. See also L. Ferry and A. Renault, *La Pensée 68. Essai sur l'anti-humanisme contemporain*, 1985.

48 Philippe Sollers, one of the founders of the journal *Tel Quel* and writer in the 1960s of novels more radically based on the play of language than those of the 'nouveau roman' – in contrast to his work of the 1980s. See, for example, *Femmes*, 1983, and *Portrait du joueur*, 1985.

49 'the arrival of Don Quixote reveals the omnipotence of desire (and the desire for omnipotence) which is the essence of the thinking of the child'; 'the man fallen from heaven . . . means to engender himself without the assistance of human parents (which doesn't prevent him from taking a passionate interest in the fictitious families whose mighty deeds are told in his books of chivalry, which is why, while repudiating his ancestry, he has a taste for genealogies)'. *Roman des origines et origines du roman*, 1972, p. 191.

Bibliography

Unless otherwise indicated, the place of publication is Paris or London.

Books by Claude Simon

Le Tricheur, Sagittaire, 1945; reissued by Minuit.
La Corde raide, Sagittaire, 1947; reissued by Minuit.
Gulliver, Calmann-Lévy, 1952.
Le Sacre du printemps, Calmann-Lévy, 1954; Livre de poche, 1975.
Le Vent, Minuit, 1957.
L'Herbe, Minuit, 1958; Minuit 'double', 1986, with an essay by A. Duncan.
La Route des Flandres, Minuit, 1960; UGE, 1963, with an essay by J. Ricardou; Minuit 'double', 1982, with an essay by L. Dällenbach.
Le Palace, Minuit, 1962; UGE, 1971; Methuen, 1972, edited and with an introduction by J. Sturrock.
Femmes, Maeght, 1966. Text by Claude Simon, 23 colour plates by Joan Miró. Reprinted, text only, in 'Claude Simon', *Entretiens*, 31, 1972, pp. 169–78; and again as *La Chevelure de Bérénice*, Minuit, 1983.
Histoire, Minuit, 1967; Folio, 1973.
La Bataille de Pharsale, Minuit, 1969.
Orion aveugle, Geneva, Skira, 'Les Sentiers de la création', 1970. Text by Claude Simon, accompanying illustrations of photographs, drawings, and paintings by Rauschenberg, Poussin, Dubuffet, Picasso and others.
Les Corps conducteurs, Minuit, 1971.
Triptyque, Minuit, 1973.
Leçon de choses, Minuit, 1976.
Les Géorgiques, Minuit, 1981.
Discours de Stockholm, Minuit, 1986. Speech on receiving the Nobel Prize.
L'Invitation, Minuit, 1987.

Album d'un amateur, Remagen-Rolandseck, Rommerskirchen, 1988. Short texts accompanying photographs by Simon.
L'Acacia, Minuit, 1989.

English translations

The Wind, trans. R. Howard, New York, George Braziller, 1986.
The Grass, trans. R. Howard, New York, George Braziller, 1986.
The Flanders Road, trans. R. Howard, with an introduction by J. Fletcher, Calder, 1985; New York, Riverrun, 1985.
The Palace, trans. R. Howard, with an introduction by J. Fletcher, Calder, 1987.
Conducting Bodies, trans. H. R. Lane, Calder and Boyars, 1975; New York, Riverrun, 1980.
Triptych, trans. H. R. Lane, Calder, 1977; New York, Riverrun, 1982.
The World About Us (*Leçon de choses*), trans. D. Weissbort, with an introduction by M. W. Andrews, Princeton, New Jersey, Ontario Review, 1983.
The Georgics, trans. B. and J. Fletcher, Calder, 1989; New York, Riverrun, 1989.
The Invitation, trans. J. Cross, with an afterword by L. Oppenheim, Dalkey Archive Press, 1992.
The Acacia, trans. R. Howard, New York, Pantheon Books, 1991.

Interviews, short texts and articles by Claude Simon

Alphant, M., 'La route du Nobel', *Libération*, 10 December 1985, pp. 27–8.
— ' "Et à quoi bon inventer?" ', *Libération*, 31 August 1989.
Bourdet, D., 'Images de Paris: Claude Simon', *La Revue de Paris*, January 1961, pp. 136–41. Reprinted in *Brèves Rencontres*, 1963, pp. 215–24.
Chapsal, M., 'Entretien avec Claude Simon', *L'Express*, 10 November 1960, pp. 30–1.
— 'Le jeune roman', *L'Express*, 12 January 1961, pp. 31–3.
— 'Entretien. Claude Simon parle', *L'Express*, 5 April 1962, pp. 32–3
Duncan, A., 'Interview with Claude Simon', in *Claude Simon: New Directions*, ed. A. Duncan, Edinburgh, Scottish Academic Press, 1985.
Duranteau, J., 'Claude Simon. "Le roman se fait, je le fais, et il me fait" ', *Les lettres françaises*, 564, 13–19 April 1967, p. 4.
Eribon, D., 'Entretien. Fragments de Claude Simon', *Libération*, 29–30 August 1981, p. 21.
Janvier, L., 'Réponses de Claude Simon à quelques questions écrites de Ludovic Janvier', *Entretiens: Claude Simon*, 31, 1972, pp. 15–29. English translation by B. Bray in *The Review of Contemporary Fiction*, 5 (1), 1985, pp. 24–33.

Juin, H., 'Les secrets d'un romancier', *Les lettres françaises*, 6–12 October 1960, p. 5.

Le Clec'k, G., 'Claude Simon, prix de la nouvelle vague: "je ne suis pas un homme orchestré" ', *Témoignage chrétien*, 16 December 1960, pp. 19–20.

Poirson, A. and Goux, J.-P., 'Un homme traversé par le travail. Entretien avec Claude Simon', *La Nouvelle Critique*, 105, June–July 1977, pp. 32–46.

Simon, C., 'Babel', *Les Lettres nouvelles*, 31, 1955, pp. 391–413.

— 'Le cheval', *Les Lettres nouvelles*, February and March 1958, pp. 168–89, 379–93.

— '*Problèmes du nouveau roman*: trois avis autorisés', *Les lettres françaises*, 11–17 October 1967, p. 13.

— 'L'opinion des romanciers' [on Ricardou's *Pour une théorie du nouveau roman*], *La Quinzaine littéraire*, 121, 1–15 July 1971, p. 10.

— 'La fiction mot à mot', in Ricardou 4 van Rossum-Guyon (eds), *Nouveau roman: hier, aujourd'hui*, vol. 2, Union générale d'éditions, 1972, pp. 73–97.

— 'Claude Simon à la question', in *Claude Simon: colloque de Cerisy*, Union générale d'éditions, 1975, pp. 403–31.

— 'La déroute des Flandres', *Le Figaro*, 13 July 1990, pp. 10-11.

— *Photographies 1937–1970*, Maeght, 1992.

Books and articles on Claude Simon

Apeldoorn, J. van, 'Une pratique de l'intertextualité: Claude Simon, lecteur d'Orwell', Chapter 3 in *Pratiques de la description*, Amsterdam, Rodopi, 1982.

— 'Claude Simon: mots, animaux et la face cachée des choses', *Cahiers de recherches interuniversitaires et néerlandaises*, 19, 1988, pp. 72–82.

Bertrand, M., *Langue romanesque et parole scripturale: essai sur Claude Simon*, Presses Universitaires de France, 1987.

Birn, R., 'Proust, Claude Simon and the Art of the Novel', *Papers on Language and Literature*, Urbana, Southern Illinois UP, 1977, pp. 168–86.

Birn, R., and Gould, K. (eds.), *Orion Blinded: Essays on Claude Simon*, Lewisburg, Bucknell UP, 1981.

Brewer, M. M. 'Narrative fission: event, history and writing in *Les Géorgiques*', *Michigan Romance Studies*, 6, 1986, pp. 27–39.

Britton, C., 'Diversity of discourse in Claude Simon's *Les Géorgiques*', *French Studies*, 38, 1984, pp. 423–42.

— *Claude Simon: Writing the Visible*, Cambridge UP, 1987.

— *Claude Simon* (ed.), Longman, 1993.

Brosman, K. S., 'Man's animal condition in *La Route des Flandres*', *Sym-*

posium, 29, 1975, pp. 57–68.

Brox Birn, R., and Budig, V., 'Deux hommes et un texte: Simon face à Rousseau, Proust et Orwell', *La Revue des sciences humaines*, 94, (220), 1990, pp. 63–78.

Carroll, D., *The Subject in Question: the Languages of Theory and the Strategies of Fiction*, Chicago, Chicago UP, 1982.

— 'Narrative poetics and the crisis in culture: Claude Simon's return to history', *L'Esprit créateur*, 28 (4), 1987, pp. 48–59.

Chalon, J., 'Enquête. Les débuts obscurs d'écrivains célèbres', *Le Figaro*, 11 March 1972, p. 14.

Châtelet, F., 'Une vision de l'histoire', *Critique*, 37, November 1981, pp. 1218–26.

'Claude Simon an der Spitze', *Bietigheimer Zeitung*, 27 November 1992.

Claudon, F., '*Histoire* et la représentation de l'histoire chez Claude Simon', *Beiträge zur Romanischen Philologie*, 26, 1987, pp. 253–8.

Dällenbach, L., *Claude Simon*, Seuil, 1988.

— '*Les Géorgiques* ou la totalisation accomplie', *Critique*, 414, 1981, pp. 1226–42.

Daprini, P., 'Claude Simon, History and "l'innommable realité" ', in *Literature and War*, ed. D. Bevan, Amsterdam, Rodopi, 1990, pp. 167–78.

Deguy, M., 'Claude Simon et la représentation', *Critique*, 187, 1962, pp. 1009–32. Trans. A. Williams as 'Claude Simon and representation' in *Claude Simon*, ed. C. Britton, Longman, 1993.

Duffy, J., 'Claude Simon, Merleau-Ponty and perception', *French Studies*, 46, 1992, pp. 33–52.

— '*Les Géorgiques* by Claude Simon: a work of synthesis and renewal', *Australian Journal of French Studies*, 21, 1984, pp. 161–79.

— '(Ms)reading Claude Simon: a partial analysis', *Forum for Modern Language Studies*, 23, 1987, pp. 228–40.

— 'The Subversion of historical representation in Claude Simon', *French Studies*, 41, 1987, pp. 421–37.

— 'Antithesis in Simon's *Le Vent*: authorial red herrings versus readerly strategies', *Modern Language Review*, 83, 1988, pp. 571–85.

Duncan, A., 'Claude Simon and William Faulkner', *Forum for Modern Language Studies*, 9, 1973, pp. 235–52.

— 'Le nouveau roman jurassien', *Le Jura français*, 58, 1991, pp. 7–11.

Ellison, D., 'Narrative levelling and performative pathos in Claude Simon's *Les Géorgiques*', *French Forum*, 12, 1987, pp. 303–21.

Evans, M., *Claude Simon and the Transgressions of Modern Art*, Macmillan, 1988.

— 'The Orpheus myth in *Les Géorgiques*', in A. Duncan (ed.), *Claude Simon: New Directions*, Edinburgh, Scottish Academic Press, 1985, pp. 89–99.

Fletcher, J., 'Intertextuality and interfictionality: *Les Géorgiques* and *Homage to Catalonia*', in A. Duncan (ed.), *Claude Simon: New Dirctions*, Edinburgh, Scottish Academic Press, 1985, pp. 100–12.

Gaudin, C., 'Niveaux de lisibilité dans *Leçon de choses* de Claude Simon', *Romanic Review*, 68, 1977, pp. 178–96.

Gosselin, C. H., 'Voices of the past in Claude Simon's *La Batalle de Pharsale*', *New York Literary Forum*, 2, 1978, pp. 23–33.

Heath, S., 'Claude Simon', Chapter 4 in *The Nouveau Roman. A Study in the Practice of Writing*, Elek, 1972, pp. 153–78.

Hollenbeck, J., *Eléments baroques dans les romans de Claude Simon*, La pensée universelle, 1982.

Janvier, L., 'Vertige et parole dans l'oeuvre de Claude Simon', in *Une parole exigeante*, Minuit, 1964, pp. 89–110.

Jost, F., 'Les aventures du lecteur', *Poétique*, 29, 1977, pp. 77–89.

Labriolle, J. de, 'De Faulkner à Claude Simon', *La Revue de littérature comparée*, 53, 1979, pp. 358–88.

Lanceraux, D., 'Modalités de la narration dans *La Route des Flandres*', *Poétique*, 14, 1973, pp. 235–49.

Lotringer, S., 'Cryptique', in J. Ricardou (ed.), *Claude Simon: colloque de Cerisy*, Union générale d'éditions, 1975, pp. 313–33.

Loubere, J. A. E., *The Novels of Claude Simon*, Ithaca, Cornell UP, 1975.

Makward, C.P., 'Aspects of bisexuality in Claude Simon's works', in Birn, R., and Gould, K., *Orion Blinded: Essays on Claude Simon*, Bucknell UP, 1981, pp. 219–35.

Merleau-Ponty, M., 'Cinq notes sur Claude Simon', *Médiations*, 4, 1961–62, pp. 5–10.

Mortier, R., 'Discontinu et rupture dans *La Bataille de Pharsale*', *Degrès*, 1 (2), 1973, pp. c1-c6.

Neumann, G., 'Claude Simon et Michelet: exemple d'intertextualité génératrice dans *Les Géorgiques*', *Australian Journal of French Studies*, 24, 1987, pp. 83–99.

Orr, M., 'Lytton Strachey: literary embellishment or functional intertext in Claude Simon's *Les Géorgiques?*', *French Studies Bulletin*, 26, spring 1988, pp. 14–17.

— 'Intertextual bridging. Across the genre divide in Claude Simon's *Les Géorgiques*', *Forum for Modern Language Studies*, 26, 1990, pp. 231–9.

— *Claude Simon: the Intertextual Dimension*, Glasgow, University of Glasgow French and German Publications, 1993.

Pingaud, B., 'Sur la route des Flandres', *Les Temps modernes*, 178, 1961, pp. 1025–37.

Poirot-Delpech, B., 'Entre deux guerres', *Le Monde*, 1 September 1989. Review of *L'Acacia*.

Pugh, A. C., 'Du *Tricheur* à Triptyque, et inversement', *Etudes littéraires*, 9 (1), avril 1976, pp. 137–60.

— *Simon: 'Histoire'*, Grant & Cutler, 1982.
— 'Defeat, May 1940: Claude Simon, Marc Bloch and the writing of disaster', *Forum for Modern Language Studies*, 22, 1985, pp. 59–70.
— 'Facing the matter of history: *Les Géorgiques*', in *Claude Simon: New Directions*, Edinburgh, Scottish Academic Press, 1985, pp. 113–30.
— 'From drawing, to painting, to text: Claude Simon's allegory of representation and reading in the prologue to *The Georgics*', *The Review of Contemporary Fiction*, 5 (1), 1985, pp. 56–70.
— 'Claude Simon: fiction and the question of autobiography', *Romance Studies*, no. 8, summer 1986, pp. 81–96.
— 'Claude Simon et la route de la référence', *La Revue des sciences humaines*, 84, (220), 1990, pp. 23–45.
Reitsma-La Brujeere, C., 'Récit et méta-récit, texte et intertexte, dans *Les Géorgiques* de Claude Simon', *French Forum*, 9, 1984, pp. 225–35.
— *Passé et présent dans 'Les Géorgiques' de Claude Simon*, Amsterdam, Rodopi, 1992.
Ricardou, J., 'Un ordre dans la débâcle', *Critique*, 163, 1960, pp. 1011–24. Reprinted in amended form in *Problèmes du nouveau roman*, Seuil, 1967, pp. 44–55.
— 'La bataille de la phrase', in *Pour une théorie du nouveau roman*, Seuil, 1971, pp. 118–58.
— 'L'essence et les sens', in *Pour une théorie du nouveau roman*, Seuil, 1971, pp, 200–10.
— 'Un tour d'écrou textuel', *Le Magazine littéraire*, 74, 1973, pp. 32–3.
— 'Le dispositif osiriaque', in *Nouveaux problèmes du roman*, Seuil, 1978, pp. 179–243.
Ricardou, J. (ed.) *Claude Simon: colloque de Cerisy*, Union générale d'éditions, 1975.
Riffaterre, M., 'Orion voyeur: l'écriture intertextuelle de Claude Simon', *Modern Language Notes*, 103, 1988, pp. 711-35.
Rinaldi, A., 'Claude Simon: un catalogue cousu de fil blanc', *L'Express*, 12–18 September 1981, pp. 72–3. Review of *Les Géorgiques*.
Roubichou, G., *Lecture de 'L'Herbe' de Claude Simon*, Lausanne, L'Age d'homme, 1976.
Sarkonak, R., *Claude Simon: les carrefours du texte*, Toronto, Paratexte, 1986.
— *Understanding Claude Simon*, Columbia, SC, South Carolina UP, 1990.
— 'Un drôle d'arbre: *L'Acacia* de Claude Simon', *Romanic Review*, 82, 1991, pp. 210–32.
Silverman, M., 'Fiction as process: the later novels of Claude Simon', in A. Duncan, *Claude Simon: New Directions*, Edinburgh, Scottish Academic Press, 1985, pp. 61–74.
Sykes, S., *Les Romans de Claude Simon*, Minuit, 1979.

— 'The Novel as conjuration: *Absalom, Absalom*! and *La Route des Flandres*', *La Revue de littérature comparée*, 53, 1979, pp. 348–57.
— '*Les Géorgiques*: "une reconversion totale"?', *Romance Studies*, 2, 1983, pp. 80–9.
Tschinka, I., 'La fabrique du corpos et la corrida du hors-corps', in J. Ricardou (ed.), *Claude Simon: colloque de Cerisy*, Union générale d'éditions, 1975, pp. 395–9.
Van Rossum-Guyon, F., 'De Claude Simon à Proust: un exemple d'intertextualité', *Marche Romane*, 21 (1–2), 1971, pp. 71–92; *Les Lettres nouvelles*, September 1972, pp. 107–37.
— 'Ut pictura poesis. Une lecture de *La Bataille de Pharsale*', *Degrès*, 1 (3), juillet 1973, pp. K1-K15.
Woodhull, W., 'Reading Claude Simon: gender, ideology, representation', 'L'Esprit créateur, 27, 1987, pp. 5–16.

General

'A l'attention des auteurs', *France–Observateur*, 13 February 1958, p. 18.
Alter, J., *La Vision du monde d'Alain Robbe-Grillet*, Geneva, Droz, 1966.
Althusser, L., *Pour Marx*, Maspéro, 1966.
Barthes, R., 'Introduction à l'analyse structurale des récits', *Communication*, 8, 1966. Reprinted in *Poétique du récit*, Seuil, 1977, pp. 7–57, and translated in *Image, Music, Text*, Fontana, 1977, pp. 79–124.
— *Système de la mode*, Seuil, 1967.
— 'La mort de l'auteur', *Mantéia*, 5, 1968. Translated as 'The death of the author', in *Image, Music, Text*, Fontana 1977, pp. 142–8.
— *S/Z*, Seuil, 1970.
— *Roland Barthes*, Seuil, 1975.
Bebbington, D., *Patterns in History. A Christian Perspective on Historical Thought*, Leicester, Apollos, 2nd edn 1990.
Belsey, C., *Critical Practice*, Methuen, 1980, Routledge, 1991.
Bloch, M., *L'Etrange défaite*, Gallimard (Folio), 1990.
Bloom, H., *The Anxiety of Influence*, Oxford UP, 1973.
Bowie, M., *Lacan*, Fontana, 1991.
Britton, C., *The Nouveau Roman: Fiction, Theory and Politics*, Macmillan, 1992.
Butor, M., *Degrés*, Gallimard, 1960.
— Répertoire, Minuit, 1960.
Dällenbach, L., *Le Récit spéculaire: essai sur la mise en abyme*, Seuil, 1977.
Dostoevski, F., *The Idiot*, trans. E. M. Martini, Dent, 1953.
Doubrovsky, S., *Fils*, 1977, Galilée.
— *Le Livre brisé*, Grasset, 1989.
Eco, U., *L'Oeuvre ouverte*, Seuil, 1965.

— *The Role of the Reader: Explorations in the Semiotics of Texts*, Hutchinson, 1981.
Faulkner, W., *Absalom, Absalom!*, Chatto & Windus, 1965.
Ferry, L., and Renaut, A., *La Pensée 68: essai sur l'anti-humanisme contemporain*, Gallimard, 1985.
Freud, S., 'Creative writers and day-dreaming', *Standard Edition of the Complete Psychological Works*, vol. 9, Hogarth, 1959, pp. 143–53.
— *On Sexuality*, Pelican Freud Library, 7, Penguin, 1977.
Genette, G., *Figures*, vols. 1, 2 and 3, Seuil, 1966, 1969, 1972.
— *Palimpsestes*, Seuil, 1982.
Henriot, E., 'Un nouveau roman', *Le Monde*, 22 May 1957.
Jakobson, R., 'Two Aspects of Language and Two Types of Aphasic Disturbance', in *Language in Literature*, Cambridge, Mass., Harvard UP, 1987, pp. 95–120. First published in *Fundamentals of Language*, The Hague, 1956.
Jefferson, A., 'Autobiography as intertext: Barthes, Sarraute, Robbe-Grillet', in *Intertextuality: theories and practices*, ed, M. Worton, and J. Still, Manchester and New York, Manchester UP, 1990, pp. 108–29.
Jenny, L., 'La stratégie de la forme', *Poétique*, 27, 1976, pp. 257–81.
Johnson, B., *The Critical Difference: Essays on the Contemporary Rhetoric of Reading*, Baltimore, Johns Hopkins UP, 1981.
Jost, F. (ed.), 'Robbe-Grillet', *Obliques*, 16–17, 1978.
Kristeva, J., 'Bakhtine, le mot, le dialogue et le roman', *Critique*, 239, 1967, pp. 438–65. Reprinted as 'Le mot, le dialogue et le roman' in *Semeiotiké: recherches pour une sémanalyse*, Seuil, 1969, pp. 143–73.
Lacan, J., *Ecrits*, Seuil, 1966.
— *Le Séminaire, livre 11: les quatre concepts fondamentaux de la psychanalyse*, Seuil, 1973.
— *Ecrits: a Selection*, trans. A. Sheridan, Tavistock Publications, 1977.
Laplanche, J., and Pontalis, J. B., *Vocabulaire de la psychanalyse*, Presses Universitaires de France, 1973.
Lejeune, P., *L'Autobiographie en France*, Colin, 1971.
— *Le Pacte autobiographique*, Seuil, 1975.
— *Lire Leiris. Autobiographie et langage*, Klincksieck, 1975.
— *Moi aussi*, Seuil, 1986.
— *La Mémoire et l'oblique: Georges Perec autobiographe*, POL, 1991.
Lentriccia, F., *After the New Criticism*, Athlone Press, 1980; Methuen, 1983.
Le Roy Ladurie, E., *Histoire du climat depuis l'an mil*, 2 vols., Flammarion, 1990.
Levinson, M., ed., *Rethinking Historicism: Critical Readings in Romantic History*, Oxford, Blackwell, 1989.
Lyotard, J.-F., *Discours, figures: un essai d'esthétique*, Klincksieck, 1978.
McLellan, D., *Karl Marx. Selected Writings*, Oxford, Oxford UP, 1977.

Merleau-Ponty, M., *Phénoménologie de la perception*, Gallimard, 1945 (1976).
Mitchell, J. (ed.), *The Selected Melanie Klein*, Penguin, 1986.
Morrissette, B., *Les Romans de Robbe-Grillet*, Minuit, 1963 and 1971.
Oppenheim, L. (ed.), *Three Decades of the French New Novel*, Urbana and Chicago, University of Illinois Press, 1986.
Pavell, T. G., *The Feud of Language: a History of Structuralist Thought*, Oxford, Blackwell, 1989.
Ricardou, J., *La Prise de Constantinople*, Minuit, 1965.
— *Problèmes du nouveau roman*, Seuil, 1967.
— *Pour une théorie du nouveau roman*, Seuil, 1971.
— *Le Nouveau Roman*, Seuil, 1973.
— *Nouveaux problèmes du roman*, Seuil, 1978.
— 'Les raisons de l'ensemble', *Conséquences*, no. 5, 1985, pp. 62–77.
Ricardou, J. (ed.), *Robbe-Grillet: analyse, théorie*, 2 vols. (Colloque de Cerisy), Union générale d'éditions, 1976.
Ricardou, J., and van Rossum-Guyon, F. (eds.), *Nouveau roman: hier, aujourd'hui*, 2 vols., Union générale d'éditions, 1972.
Riffaterre, M., 'Compulsory reader response: the intertextual drive', in M. Worton and J. Still (eds), *Intertextuality: theories and practices*, Manchester University Press, 1990, pp. 56–76.
Robbe-Grillet, A., 'Le réalisme, la psychologie et l'avenir du roman', *Critique*, 111–12, 1956, pp. 695–701.
— 'Une voie pour le roman futur', *La Nouvelle Revue française*, July 1956, pp. 77–84.
— *Pour un nouveau roman*, Gallimard (Idées), 1963.
— 'Order and disorder in film and fiction', *Critical Enquiry*, autumn 1977, pp. 1–20.
— 'Fragment autobiographique imaginaire', *Minuit*, 31, 1978, pp. 2–8.
— *Le Miroir qui revient*, Minuit, 1985.
— *Angélique ou l'enchantement*, Minuit, 1988.
Robert, M., *Roman des origines et origines du roman*, Grasset, 1972.
Rousseau, J.-J., *Les Confessions*, Garnier-Flammarion, 1968.
Sarraute, N., *L'Ere du soupçon*, Gallimard, 1956.
— *Enfance*, Gallimard, 1981.
Sartre, J. P., *L'Imaginaire*, Gallimard, 1940.
— *Qu'est-ce que la littérature?*, Gallimard (Idées), 1948.
— *Les Mots*, Gallimard, 1964.
Sollers, P., *Femmes*, Gallimard, 1983.
— *Portrait du joueur*, Gallimard, 1985.
Sturrock, J., *The French New Novel*, Oxford UP, 1969.
Tel Quel. Théorie d'ensemble, Seuil, 1968.
Todorov, T., *Qu'est-ce que le structuralisme?, 2. Poétique*, Seuil, 1968.
Vidal, P., *Histoire de la révolution française dans tes Pyrénées orientales*, 3

vols., Perpignan, Imprimerie de l'Indépendant, 1885, 1886, 1889.
Worton, M. and Still, J. (eds.), *Intertextuality: theories and practices*, Manchester, Manchester UP, 1990.

Index